S0-CAS-658

The Organic Chemistry of Palladium

Volume II CATALYTIC REACTIONS

ORGANOMETALLIC CHEMISTRY

A Series of Monographs

BRIAN G. RAMSEY: Electronic Transitions in Organometalloids, 1969.
R. C. POLLER: The Chemistry of Organotin Compounds, 1970.
RUSSELL N. GRIMES: Carboranes, 1970.
PETER M. MAITLIS: The Organic Chemistry of Palladium, Volume I – 1971, Volume II – 1971.

The Organic Chemistry of Palladium

Peter M. Maitlis
McMaster University
Hamilton, Ontario, Canada

Volume II ***Catalytic Reactions***

1971 Academic Press New York and London

ACADEMIC PRESS, INC.
111 Fifth Avenue, New York, New York 10003

United Kingdom Edition published by
ACADEMIC PRESS, INC. (LONDON) LTD.
Berkeley Square House, London W1X 6BA

LIBRARY OF CONGRESS CATALOG CARD NUMBER: 77-162937

PRINTED IN THE UNITED STATES OF AMERICA

E quindi
uscimmo
a riveder
le stelle
—Dante

To Marion, Niccola, Sally, and Emily

Contents

Preface

The last twenty years have seen the emergence of the new field of organo-transition metal chemistry. It has now become one of the most important and exciting areas of chemical endeavor. This stems not only from the use of organo-transition metal complexes as catalysts and models for catalytic reactions, but also from their intrinsic importance to our understanding of structure and bonding.

Only a few years ago our knowledge of the field could be comfortably fitted into one volume; today the same space would hardly suffice to cover one of the more popular aspects of the subject, and any attempt to deal comprehensively with it therefore requires some subdivision. Useful lessons can be learned from a classification according to either the metal, the ligand, or the reaction, but each such arrangement has its drawbacks. In the light of current knowledge a treatment based on the metal is perhaps the most rewarding, especially if this is accompanied by references to other, neighboring metals, and further subdivision by ligands and reactions.

The choice of palladium for the first work on the Transition Metals in The Organometallic Chemistry Series was dictated to a large degree by the number of organic reactions which palladium catalyzes. These include olefin oxidation, the oligomerization of olefins, dienes and acetylenes, carbonylation, coupling of arenes, vinylation, acetoxylation, isomerization, halogenation, and many others. In addition, the chemistry of the organometallic complexes of palladium and platinum is well explored and, in outline at least, well understood.

A complete treatment of even this limited topic covers a great deal of material and a wide variety of reactions. I have accordingly chosen to divide the monograph into two volumes, the first covering the metal complexes and their structures, bonding, and reactions, while the second deals with the catalytic processes and other reactions induced by palladium. Although I have treated the entire topic in a unified fashion, each volume is self-contained and may be read separately.

The catalytic reactions are of such variety that palladium promises to become as important in organic synthesis as the Grignard reagent or hydroboration, and may well be more versatile than either. In this connection, a most significant feature of palladium chemistry is the ease of reoxidation of Pd(0) to Pd(II). This has allowed the active Pd(II) to be regenerated *in situ* and has made the industrial synthesis of acetaldehyde from ethylene (Wacker process) under homogeneous conditions not only feasible but economically more attractive than any other route. This use of a rather rare and expensive metal in an industrial process only foreshadows other equally important developments.

Even when the commercial exploitation of a *homogeneously* catalyzed reaction appears to be unfavorable, the study of such processes can lead to the development of heterogeneous catalysts of high specificity for unusual reactions. The one-step palladium(II)-catalyzed homogeneous synthesis of vinyl acetate from ethylene is now, as a heterogeneous reaction, the most economical method for the large-scale production of vinyl acetate. While this is likely to be the pattern of much future industrial use of the platinum metals, only studies under homogeneous conditions can lead to a detailed understanding of the reactions involved and to rational new developments.

The aim of this two-volume work has been to collect the available data, both on the complexes and on the catalyzed reactions, and to fit them into a coherent pattern. A number of mechanisms are well established, but in many cases one can only speculate on reaction paths. In order to facilitate the development of rational hypotheses, I have included information on the inorganic chemistry of palladium and also on the chemistry, where it is relevant, of the neighboring elements. By doing so, I hope also to have defined more exactly the unique features of the metal. On present evidence we can conclude that the special features exhibited by Pd(II), for example, arise from its greater lability by comparison with Pt(II) and the lower affinity toward oxygen and nitrogen donor ligands by comparison with the even more labile Ni(II). Of particularly great interest is the question of whether, by a suitable choice of ligand, one metal cannot be "tuned" in such a way that it chemically resembles another. If this is possible, and present indications are that it is, then it should also be feasible to carry out catalytic reactions characteristic of palladium with other metals.

Although this work is designed specifically for the research worker in the field and for the organic chemist who wishes to make use of the wide variety of metal-catalyzed reactions now available to him, the inclusion of comparative studies and the introductory sections should also make it useful as a supplementary text for graduate courses.

Since I have tried to make these volumes of particular use to the organic chemist, I have emphasized the broader aspects of the mechanisms and have not included extensive tabulations of physical properties of complexes, which may all be found in the appropriate references. The newcomer to the field should, however, be warned that many of the properties of a given complex, such as the melting or decomposition points and even the color, may vary within quite wide limits.

Inorganic chemists, in particular, may be dismayed at my decision not to use rigorous IUPAC nomenclature for the complexes. In practice, this is cumbersome and tends to emphasize the metal, or other trivial ligands, at the expense of the organic ligand of interest, with the result that even a reader familiar with the field may take several minutes to grasp the structure. The simplifications used, which are essentially of such a nature as to emphasize the organic ligand [e.g., allylpalladium chloride dimer in place of di-μ-chlorobis(π-allyl)dipalladium], together with diagrammatic formulas will enable the text to be read more easily.

This work covers the literature on palladium comprehensively to 1970. In addition, a large number of references to work published in 1970 are included, and I hope that no significant advance in this area which appeared in press before 1971 has been omitted. Coverage of the chemistry of the neighboring elements and of some aspects of the inorganic chemistry of palladium is, of necessity, more curtailed.

This work was started in 1968–1969 at Imperial College, London. I should like to express my appreciation to Professor G. Wilkinson and his colleagues for their generous hospitality and to the National Research Council of Canada for the award of a Senior Fellowship which made these volumes possible.

I should also like to thank all who so kindly read parts of the manuscript for their comments, in particular, Dr. R. F. Heck, Dr. P. Henry, Dr. J. Powell, Dr. J. F. Harrod, Dr. J. M. Davidson, and Professor G. C. Bond.

Peter M. Maitlis

Contents of Volume I

Metal Complexes

The Organic Chemistry of Palladium

Volume II CATALYTIC REACTIONS

Introduction

The chapters which make up Volume II of this monograph are concerned with organic transformations caused by palladium complexes (or metal) either stoichiometrically or catalytically. Many of these reactions are of considerable interest and importance, both synthetically and industrially.

Elucidation of the mechanisms of these reactions is therefore of great importance. Unfortunately, the experimental evidence is lacking to a considerable extent. This is not to imply that much good work has not been done in the field, but merely that the processes involved are very complex, considerably more so, in fact, than "normal" organic transformations. Thus, even a careful kinetic study of a reaction will only rarely, by itself, give definitive evidence for a reaction path. The normal kinetic analysis only gives information about the slowest, rate-determining step. In many cases this is relatively trivial and may be, for example, the rate of π-complex formation. The important question of what happens in the coordination sphere once the complex has been formed remains unanswered since these reactions are often very fast. An example of this is in the oxidation or isomerization of olefins catalyzed by Pd(II) where the key step, a hydride transfer, is only very poorly understood.

Another problem is that many of the metal-catalyzed reactions give rise to a range of products. Particularly in the oxidation reactions, a complete and accurate materials balance is very hard to achieve. This problem adds to the difficulty of disentangling the processes involved.

Despite such reservations considerable progress has been made, especially in recent years, in our understanding of metal-catalyzed reactions. A very useful method for analyzing a given reaction is to use model systems, frequently with a different metal atom. Thus rhodium and palladium are very active in homogeneously catalyzed reactions, whereas iridium and platinum on the whole react much more slowly and therefore lend themselves to a step-by-step analysis. Bearing in mind the limitations of all arguments by analogy, it is now possible to propose mechanisms for a variety of catalytic processes.

The typical mechanism involves the following steps:

(1) Oxidative addition of a molecule A–B to a metal complex $[L_mM^{n+}]$ to give $[L_pM^{(n+2)+}AB]$.

(2) Creation of a vacant coordination site on the metal by loss of one ligand, L, followed by coordination of a further molecule Y (frequently an olefin) to give $[L_{p-1}M^{(n+2)+}ABY]$.

(3) Rearrangement of the groups A, B, and Y in the complex to give the intermediate in which these groups are so placed as to facilitate reaction.

(4) Insertion of Y into M–A (or M–B) giving $[L_{p-1}M^{(n+2)+}(YA)B]$; this is frequently a solvent-assisted process.

(5) Regeneration of the catalyst, $[L_mM^{n+}]$, with loss of AY, either by reaction with B (to give AYB), or by a rearrangement (e.g., a hydride shift) to give a stable organic product, A′Y′.

In many cases these steps are modified and, even when the reaction is quite well understood, it is often difficult to be too exact about the nature of the "catalyst". In the case of the palladium-catalyzed oxidations the reactive entity, in the above sense, is Pd(0) which is oxidized to Pd(II). However, since Pd(II) compounds are readily available and very reactive, the oxidative addition step is usually carried out at the end of the cycle. Normally such oxidation reactions are stoichiometric and result in the formation of inactive palladium metal. As discussed in Chapter II, however, Pd(0) is very easily reoxidized to Pd(II); Cu(II) is a very popular oxidant for this reaction since it can again be regenerated with oxygen. For this reason palladium catalyzed oxidation reactions are synthetically and industrially attractive.

Palladium(II) also catalyzes a large number of reactions which do not involve reduction to Pd(0); some of these occur by insertion into Pd–X bonds. It is also possible that some involve Pd(IV) intermediates, but there is little evidence for this at the moment.

A curious feature of reactions catalyzed by transition metals is the relative lack of evidence for the typical reactive intermediates of organic chemistry, carbanions and carbonium ions. This probably arises from the metal acting both as a source and a sink of electrons and making such ionic intermediates energetically unfavorable. Although organic free radical intermediates have been postulated as intermediates in metal-catalyzed reactions, there is again little evidence for them. Most reactions should, therefore, on present evidence, be regarded as occurring largely within the coordination sphere of the metal. Since the most kinetically labile metal complexes are usually the most reactive this also accounts for the high rates of palladium-catalyzed reactions.

A problem which has not, as yet, been resolved, is the coordination number and stereochemistry of the various intermediates involved. Palladium(II) is known to favor square planar four-coordination and most mechanisms proposed assume that the intermediates also have this coordination. This is very probably an oversimplification since five-coordinate complexes are known to exist and have long been proposed as transition states in substitution reactions.

Most homogeneously catalyzed reactions which have so far been investigated appear to involve only one metal atom. This is in contrast to the heterogeneously catalyzed reactions which occur on metal surfaces, and where the participation of two or more metal atoms is usually invoked. While this appears to be generally true for the reactions of Pd(II), there is some evidence that, under certain conditions, two or more metal atoms can also participate in homogeneous reactions. The extent to which this occurs and the nature of these reactions remain unknown.

Chapter I

The Formation and Cleavage of Carbon–Carbon Bonds

As has been discussed in the first Volume, nucleophilic attack on a carbon atom of a π-bonded ligand is a favored reaction path for complexes of Pd formally in the (II) oxidation state. Of particular interest are those reactions in which a C–C bond results. Only in a few cases is the nucleophile clearly distinguished as such; nevertheless these reactions will all be grouped together for convenience. This does *not* imply any similarity of reaction path among them.

For present purposes the reactions will be discussed under the following headings:

A. Addition of a carbanionoid (organometallic) species (e.g., malonate, Grignard, alkyl- or arylpalladium, organomercury) to an olefin to give either a new complex or an organic product.

B. Carbonylation, again both of complexes and (formally) uncomplexed organic species.

C. Coupling of olefins (and dienes), either with themselves or with other organic moieties in which no net oxidation of the organic ligand (or reduction of the metal) occurs. These reactions usually involve a H shift. Most acetylene oligomerization reactions are also nonoxidative and only cases where H shifts do not occur are treated.

D. Coupling of organic molecules with loss of H (usually one per molecule of substrate) and consequent reduction of Pd(II) to Pd(0). These reactions are very important; since the Pd(0) can be reoxidized to Pd(II), they are potentially catalytic.

E. Miscellaneous.

F. Cleavage of C–C bonds, including decarbonylation reactions and skeletal rearrangements.

A. ADDITION OF ORGANOMETALLIC (AND SIMILAR CARBANIONOID) SPECIES TO OLEFINS

Some of these reactions on π complexes lead to new complexes and have already been discussed in some detail elsewhere (for example, Volume I, Chapter II, Section C,3,b,iii).

Perhaps the simplest examples are the addition of malonate, acetoacetate, etc., to 1,5-cyclooctadienepalladium chloride described by Tsuji *et al.*[1-3] Thallium salts of β-diketonates have been used similarly by Johnson *et al.*[4] The most interesting point here is that the product always has the substituent exo to the metal and contains a metal–carbon σ bond. This implies, assuming no gross overall rearrangement, that attack is on the olefin (exo) and that the metal does not participate directly.

$$(C_8H_{12})PdCl_2 + CHXY^- \longrightarrow \tfrac{1}{2}\left[(C_8H_{12}CHXY)PdCl\right]_2 + Cl^-$$

Reactions of this type can also proceed further with "insertion" of the coordinated olefin into the Pd–C σ bond.† This can occur under the action of nucleophiles[5] or light[6] (Volume I, Chapter III, Section E,3; Volume I, Chapter IV, Section F,1,c).

† The term "insertion" is used in a nonspecific sense; it is probable that reaction proceeds by *migration* of R (in R–M) onto the coordinated olefin, but insufficient data are available for it to be certain that this mode of reaction applies in all cases (see Section B below).

H OMe PdCl 2 + diphos ⟶ OMe H Ph_2P—Pd—Cl PPh_2

OR PdOAc 2 $\xrightarrow{h\nu}$ H OR

Tsuji and Takahashi[1-3] have also shown that strong bases will react with complexes of this type to give bicyclic products. Both appear to involve nucleophilic displacement of palladium by $XC(COOEt)_2^-$, either before or after insertion of the olefin into the Pd–C σ bond has occurred.

$CH(COOEt)_2$ PdCl 2 $\xrightarrow{MeSOCH_2^-/DMSO}$ $C(COOEt)_2$

$\xrightarrow{CH(COOEt)_2^-/DMSO}$ $CH(COOEt)_2$ $CH(COOEt)_2$

Shier[7] and Heck[8,9] have both reported attack (by methyl Grignard and phenylmercury, respectively) on uncoordinated dienes in the presence of a palladium(II) salt to give π-allylic complexes. In the *absence* of R–M, these

$$R\text{–}M + Pd(II) + CH_2{=}CHCH{=}CH_2 \longrightarrow \text{(}\pi\text{-allyl)—Pd(II)} \quad CH_2R$$

reactions proceed as shown.

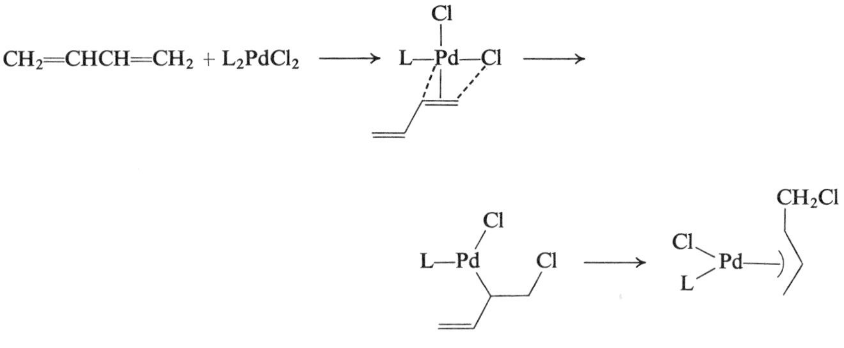

$$CH_2{=}CHCH{=}CH_2 + L_2PdCl_2 \longrightarrow$$

$$\longrightarrow$$

(L = Cl⁻, solvent)

In the presence of an alkylating or arylating reagent (at low temperatures) an R–Pd species is presumably formed first. It is known that many σ-bonded R–M complexes insert olefins very readily and palladium is no exception; in contrast, insertion into M–X, where X is halogen or an oxy group, is generally slower. A reasonable reaction sequence for the formation of 1-substituted π-allylic complexes from butadiene is, therefore,

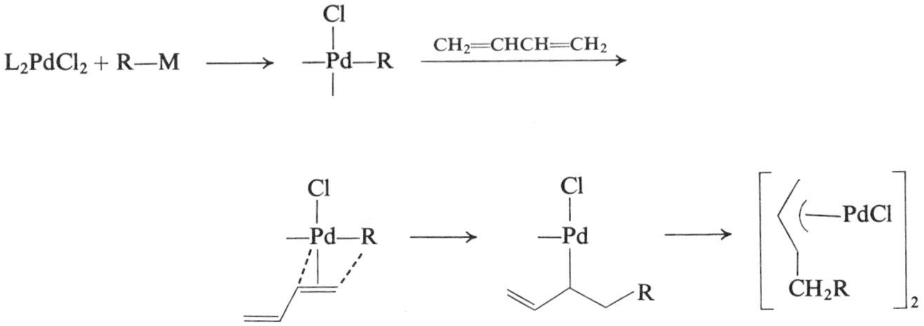

$$L_2PdCl_2 + R{-}M \longrightarrow \text{—Pd(Cl)—R} \xrightarrow{CH_2{=}CHCH{=}CH_2}$$

(R = aryl, methyl)

A mechanism involving 1,4-addition of Pd–R is also possible. A similar rationale can be given for the reaction of allene, methyl Grignard, and Li_2PdBr_4 to give 2-methylallylpalladium bromide, described by Shier.

If the allylic group in a palladium complex reacts via the σ-allyl form in the intermediate, then the reaction of such complexes with dienes can easily be understood.[10–14] Two modes of addition are found, depending on whether the diene is 1,2 (allene) or 1,3.[10]

$$\left[R'\text{—}\langle\!\langle\!\text{—PdX} \atop R \right]_2 \xrightarrow{CH_2{=}C{=}CH_2} \left[RCH{=}CR'\cdot CH_2\text{—}\langle\!\langle\!\text{—PdX} \right]_2$$

$$\left[R'\text{—}\langle\!\langle\!\text{—PdX} \atop R \right]_2 \xrightarrow{CH_2{=}CHCH{=}CH_2} \left[\langle\!\langle\!\text{—PdX} \atop CH_2CHR\cdot CR'{=}CH_2 \right]_2$$

These reactions are discussed in Section C,3,b and c, the Appendix, and in Volume I, Chapter V, Section F,12.

Tsuji *et al.*[15,16] have also reported reactions of π-allylpalladium chloride with malonate (or acetylacetonate) to give mono- and diallylmalonate (or acetylacetonate).

$$\left[\langle\!\langle\!\text{—PdCl} \right]_2 + CH_2XY \xrightarrow[(-HCl)]{\text{base}} CH_2{=}CHCH_2CHXY + (CH_2{=}CHCH_2)_2CXY + Pd$$

$$(X = COOEt;\ Y = COOEt \text{ or } COMe)$$

Similarly, 1-methylallylpalladium chloride dimer reacted with diphenylmercury to give 1-phenyl-2-butene.[9]

$$\left[\langle\!\langle\!\text{—PdCl} \atop Me \right]_2 + Ph_2Hg \longrightarrow PhCH_2CH{=}CHMe + Pd$$

Okada and Hashimoto[17] and Saegusa *et al.*[18] have described the reactions of Grignard and organomercury compounds with π-olefin complexes. Thus the styrene–palladium chloride complex gave *trans*-stilbene (R = Ph) and *trans*-β-methylstyrene (R = Me) with phenyl- and methylmagnesium bromide, respectively.

$$\left[\begin{matrix} CH_2 \\ \| \text{—PdCl}_2 \\ PhCH \end{matrix} \right]_2 \xrightarrow{RMgBr} \begin{matrix} Ph & & H \\ & C{=}C & \\ H & & R \end{matrix}$$

[β-Methylstyrene$PdCl_2$]$_2$ did not react with methyl Grignard, but [propylene-$PdCl_2$]$_2$ gave largely β-methylstyrene together with some α-methylstyrene, allylbenzene, propylbenzene, and isopropylbenzene with PhMgBr. Okada and Hashimoto suggested that the reaction proceeds via initial formation of a Pd–R′ species followed by insertion of the π-bonded olefin into Pd–R′, to give **(I-1)**.

$$\left[\text{RCH=CH}_2\text{—PdCl}_2\right]_2 \xrightarrow{\text{R'MgX}} \text{RCH=CH}_2\cdots\text{Pd(R')—Cl} \longrightarrow$$

$$\underset{\textbf{(I-1)}}{\text{RCH(PdCl)—CH}_2\text{R'}} \longrightarrow \text{RHC=CHR'} + \text{HCl} + \text{Pd}$$

These authors suggested an α elimination of H–Pd from **(I-1)** together with a hydride shift to give RCH=CHR′, but a direct β elimination of H–PdCl as shown would seem more probable. It is also conceivable that **(I-1)** can arise directly by attack of R . . . M on the olefin–metal complex. This possibility cannot be ruled out from the observed stereochemistry of the product, which may, in any case, be largely determined by thermodynamic factors.

From the variety of by-products obtained from the reaction of [propylene-$PdCl_2$]$_2$ and PhMgBr, it is apparent that attack at the substituted carbon, together with complex H transfer and isomerization processes, can also occur. It is also interesting, and probably significant, that these authors reported *lower* yields from reaction with the olefin complex than from a mixture of olefin and $PdCl_2$.

Saegusa *et al.*[18] have reported that carbomethoxylation of [cyclohexene-$PdCl_2$]$_2$ with MeOOCHgCl gave methyl cyclohexanecarboxylate in low yield. Since full details, including a complete materials balance, are unfortunately lacking, it would be inappropriate to speculate on the mechanism.

$$\left[\text{C}_6\text{H}_{10}\text{—PdCl}_2\right]_2 + \text{MeOOCHgCl} \longrightarrow \underset{(10\%)}{\text{C}_6\text{H}_{11}\text{COOMe}}$$

A series of studies by Heck has greatly added to our knowledge of these types of reactions. The first paper[19] describes a reaction basically similar to those discussed above, except that the free olefins, rather than their complexes, are used as starting materials. The carbanionoid moiety is derived from less reactive organometallics, such as those of mercury (and sometimes tin and lead). These have the advantage over Grignards in that functional groups and simpler conditions are permissible. For example, air and moisture do not interfere and reactions can, in fact, be carried out in protonic solvents (acetic acid, methanol).

$$RHgX + PdX_2 + R'CH{=}CH_2 \rightarrow R'CH{=}CHR + HX + Pd + HgX_2$$

The structures of the reactive intermediates are not known, but it is presumed that they are solvated species of the type $RPdX(solv.)_2$. For simplicity, they are referred to as RPdX in this discussion.

Organomercury compounds, RHgX (where X = Cl or OAc) and R_2Hg, were used; the palladium was usually present as Li_2PdCl_4, although palladium nitrate and acetate were also successfully utilized. Isomerization of the olefin-starting material and the product was frequently encountered in reaction mixtures containing large amounts of chloride ion. This was suppressed by use of RHgOAc in the presence of palladium acetate. Heck found that electron-releasing substituents (*p*-MeO, *p*-HO, *p*-Et_2N) on the aryl group (in ArHgX) decreased yields of product and good coordinating groups (NH_2) retarded or even stopped the reactions, owing to the formation of stable Pd complexes,

Heck showed by competition experiments with PhHgCl and $LiPdCl_3$† in MeCN that the relative rates of reaction were ethylene (14000) > vinyl acetate (970) > propylene (220) > styrene (42) > α-methylstyrene (1). He concluded that these reactions were therefore neither true radical nor ionic reactions. The size of the substituent on the olefin appeared to be of greatest importance; however, in the arylation reactions bulky groups on the aryl about the reactive site did not influence the course of reaction significantly, for example,

Me, HgOAc, Me, Me, HgOAc (1,4-bis(acetoxymercuri)-2,3,5-trimethylbenzene) $+ CH_2{=}CHCOOMe + LiPdCl_3 \xrightarrow{MeCN}$ Me, CH=CHCOOMe, Me, Me, CH=CHCOOMe

Alkyl groups without a β hydrogen and aryl groups added mostly, or, in some cases, exclusively to the less substituted carbon of an unsymmetrical olefin, and mixtures of isomers (other than cis, trans) were not obtained. The more heavily tri- and tetrasubstituted olefins reacted relatively slowly;

† $PdCl_2$ in acetonitrile only reacted with one mole of LiCl[8]; the structure of the species present in solution is not known, but it may be $Li_2Pd_2Cl_6$ or $LiPdCl_3MeCN$.

decomposition of the arylating agent to biaryl then became the major reaction path.

Some representative yields of methyl cinnamate obtained using a variety of organometallics as phenylating agents under different conditions are listed in Table I-1.

TABLE I-1

$$Ph_nM + CH_2{=}CHCOOMe \rightarrow PhCH{=}CHCOOMe^a$$

Reactants	Solvent	Yield (%)
$Ph_2Hg + LiPdCl_3$	MeCN	88
$PhHgCl + LiPdCl_3$	MeCN	53
$Ph_4Sn + Li_2PdCl_4$	MeOH	100
$Ph_4Pb + Li_2PdCl_4$	MeOH	82
$Ph_2Hg + Pd(NO_3)_2$	MeOH	83
$PhMgBr + LiPdCl_2$	THF/MeCN	8

[a] After Heck.[19]

These reactions could also be run with catalytic amounts of the palladium complex, which was regenerated by oxidation with Cu(II) or $Hg(OAc)_2$.

$$CH_2{=}CHCH_2OEt + PhHgOAc + Pd(OAc)_2 + Hg(OAc)_2 \xrightarrow{Me_2CO/20^\circ} \underset{\text{(approx. 360\% based on Pd)}}{PhCH{=}CHCH_2OEt}$$

Heck suggested that the reactive species in these solutions was a solvated aryl (or alkyl) palladium species (formed reversibly) which then reacted directly with the olefin to give aryl- (or alkyl-) ethylpalladium complexes [analogous to (**I-1**)]. The latter decomposed by β elimination of H–PdX. The hydridic species did not appear able to hydrogenate the olefin, but decomposed to metal and HX.

In the presence of a high concentration of cupric chloride, the formation of a 2-arylethyl chloride began to be the dominant reaction.[20]

$$PhHgCl + C_2H_4 + Li_2PdCl_4 + CuCl_2 + LiCl \xrightarrow{HOAc/H_2O} \underset{(76\%)}{PhCH_2CH_2Cl} + \underset{(2\%)}{PhCH{=}CH_2}$$

The bromides could be obtained similarly, though in lower yields. Increasing substitution at the olefinic carbons favored formation of the arylated olefin.

The mechanism clearly involves $CuCl_2$ and probably an alkyl–Cu compound since Heck found that even in the absence of olefin and a Pd(II) salt, benzyl-mercuric chloride reacted with $CuCl_2$ to give largely benzyl chloride. Alkylating agents must react with $CuCl_2$ much more readily than arylating agents; otherwise aryl halides would have been the products of the above reactions. One possibility is that the $[ArCH_2CH_2PdX]$ intermediate postulated before, reacts with $CuCl_2$ to give $ArCH_2CH_2CuCl$ (and PdClX), which then gives rise to $ArCH_2CH_2Cl$ and Cu(0).

Heck has also described a number of variants on the basic reaction. For example, when primary or secondary allylic alcohols are used as substrates under the above arylating conditions, 3-arylaldehydes or 3-arylketones result.†[22] This reaction is envisaged as proceeding similarly to that described

$$PhHgCl + CH_2{=}CHCH(R)OH + LiPdCl_3 + CuCl_2 \xrightarrow{MeCN} PhCH_2CH_2COR$$
$$(R = H, Me, Et)$$

above to give (**I-2**), which then eliminates HPdX preferentially to give the alcohol, $ArCH_2CH{=}CROH$, the enol form of the major product, $ArCH_2CH_2$ COR.

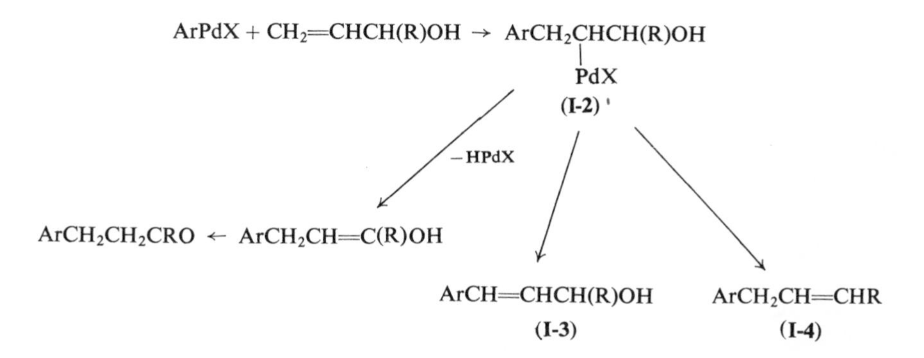

The alternative pathway, involving elimination of H from the benzylic methylene group, gives a cinnamyl alcohol (**I-3**). This pathway is followed to the extent of about 10% of that leading to the ketone. The allyl benzene (**I-4**) is also observed in some reactions, and could possibly arise from (**I-2**) by elimination of HOPdCl. However, Heck does not believe this to be the route by which (**I-4**) arises since the product is absent in chloride-free media. A more probable source is the allylic chloride (see below).

Using crotyl alcohol, some 2-aryl- as well as 3-arylaldehyde is obtained. This suggests that Ar–Pd addition to the double bond can occur in both senses.

† A sterically hindered base, dicyclohexylethylamine, was often added to the catalytic reactions in order to remove HCl.

$$ArPdX + MeCH{=}CHCH_2OH \rightarrow ArCH(Me){-}CH(PdX){-}CH_2{-}OH + MeCH(PdX){-}CH(Ar)CH_2OH$$

$$ArCH(Me)CH_2CHO \leftarrow ArCH(Me)CH{=}CHOH \leftarrow ArCH(Me){-}CH(PdX){-}CH_2{-}OH$$

$$MeCH(PdX){-}CH(Ar)CH_2OH \xrightarrow{\text{H shift}} MeCH_2C(Ar)(PdX)CH_2OH$$

$$MeCH_2C(Ar)CHO \leftarrow MeCH_2C(Ar){=}CHOH \leftarrow MeCH_2C(Ar)(PdX)CH_2OH$$

When allylic halides were arylated, allylaromatic compounds were produced.[23]

$$PhHgCl + CH_2{=}CHCH_2Cl + LiPdCl_3 \xrightarrow{MeCN/20^\circ} PhCH_2CH{=}CH_2 \quad (61\%)$$

This reaction was catalytic with respect to the palladium salt, but under some conditions isomerization of the allylbenzene to propenylbenzene was observed with concomitant precipitation of palladium. This did not generally occur in the presence of $CuCl_2$ and at high enough concentration of the allylic halide.

A mechanism again involving addition of Ar–PdX to the allylic halide, followed this time by elmination of PdX_2, was proposed. There was no evidence for products arising from elimination of HPdX, but a full investigation into

$$ArPdX + CH_2{=}CHCH_2X \rightarrow ArCH_2CH(PdX)CH_2X \rightarrow ArCH_2CH{=}CH_2 + PdX_2$$

this reaction was hampered by the isomerization to the propenylbenzene.

Vinyl acetate was phenylated under the above conditions to give *trans*-stilbene, phenylacetaldehyde, and β-acetoxystilbene, in addition to a little styrene.[24]

$$PhHgCl + CH_2{=}CHOAc + Li_2PdCl_4 + CuCl_2 \xrightarrow{HOAc}$$
$$\textit{trans}\text{-}PhCH{=}CHPh + PhCH_2CHO + PhCH{=}CHOAc + PhCH{=}CH_2$$
$$(34\%) \qquad (33\%) \qquad (30\%) \qquad (3\%)$$

Isopropenyl acetate gave benzyl methyl ketone (and chlorobenzene).

$$PhHgCl + CH_2{=}CMeOAc + Li_2PdCl_4 + CuCl_2 \xrightarrow{HOAc/H_2O} PhCH_2COMe + PhCl$$
$$(50\%) \qquad (40\%)$$

The multiplicity of products obtained in the former reaction was explained by postulating that addition of Ar–Pd could occur in both senses. The inter-

$$ArPdX + CH_2{=}CROAc \rightarrow \underset{\textbf{(I-5)}}{ArCH_2\underset{|}{C}(PdX)ROAc} + \underset{\textbf{(I-6)}}{CH_2(PdX){-}CR(Ar)OAc}$$

mediate **(I-5)** could lose either CH_3COPdX or HPdX, to give the benzyl ketone (aldehyde)† or β-acetoxystyrene, respectively, while **(I-6)** could lose PdX(OAc) to give the styrene. Stilbenes arose from the styrene, $ArCR{=}CH_2$,

$$\textbf{(I-5)} \xrightarrow{-AcPdX} ArCH_2COR$$

$$\textbf{(I-5)} \xrightarrow{-HPdX} ArCH{=}C(R)OAc$$

$$\textbf{(I-6)} \xrightarrow{-Pd(OAc)X} ArCR{=}CH_2$$

since this was more reactive than the vinylic acetate.

Vinylic halides gave the stilbenes only.[24]

Heck[9] also reported that butadiene reacted with phenylmercuric acetate in the presence of lead tetraacetate and palladium acetate to give 1-phenyl-3-butene-2-yl acetate. Similarly, monoolefins gave saturated acetates.

$$PhHgOAc + CH_2{=}CHCH{=}CH_2 + Pb(OAc)_4 \xrightarrow{Pd(OAc)_2} PhCH_2CH(OAc)CH{=}CH_2$$

$$PhHgOAc + RCH{=}CH_2 + Pb(OAc)_4 \xrightarrow{Pd(OAc)_2} PhCH_2CH(R)OAc + PhCH{=}CHR$$

$$(R = H, Me)$$

The acetates may arise from intermediates such as $[PhCH_2CHRPdX]$ by acetoxylation of the Pd–C bond.

Heck has also carried out mechanistic studies of the phenylation reaction using "phenylpalladium acetate" [from $Pd(OAc)_2$ and PhHgOAc] to avoid the isomerization which occurs in the presence of chloride ion.[25] Toward propylene in methanol, anti-Markovnikov addition (of Ph–Pd) was preferred

† Heck[21] has recently shown that $CuCl_2$ is responsible for the formation of $PhCH_2CHO$ in the phenylation of vinyl acetate. Presumably, elimination of $CH_3COPdCl$ does not occur.

6:1 over Markovnikov addition; similar results were obtained in other solvents.† Product analysis also indicated that elimination of a benzylic hydrogen was preferred 6:1 over elimination of a terminal hydrogen. Formation of trans olefin was also preferred ca. 6:1 over cis.

$$
\begin{array}{l}
CH_2{=}CHCH_3 + PhPdOAc \xrightarrow{\text{anti-Markovnikov}} PhCH_2CH(CH_3){-}Pd{-} \\
\quad \xrightarrow{(-\text{benzylic H})} \text{Ph(H)C=C(H)Me (60\%)} + \text{Ph(H)C=C(Me)H (9\%)} \\
\quad \xrightarrow{(-\text{terminal H})} PhCH_2CH{=}CH_2 \text{ (15\%)} \\
CH_2{=}CHCH_3 + PhPdOAc \xrightarrow{\text{Markovnikov}} {-}Pd{-}CH_2CH(Ph)CH_3 \longrightarrow CH_2{=}C(Ph)Me \text{ (16\%)}
\end{array}
$$

More detailed information was obtained from a study of the phenylation of the three isomeric methylstyrenes; these results are summarized in Table I-2. In all cases addition of Ph–Pd in the Markovnikov sense was negligible. Since the order of olefin reactivities reported previously gives no support for ionic or radical intermediates, the simplest explanation of these results is that cis addition (of Ph–Pd) and cis elimination (of H–Pd) predominate. These reactions are summarized in Scheme I-1 on p. 16.

If these were the only mechanisms operating, *trans*-β-methylstyrene **(I-7)**, should give rise only to *trans*-1,2-diphenylpropene, **(I-8)**, and *cis*-β-methylstyrene, **(I-11)**, to *cis*-1,2-diphenylpropene, **(I-12)**. The isolation of some of the other isomer and of the unexpected 2,3-diphenylpropene **(I-10)** in each case arises from the ability of the intermediate π-complexed hydrides (A) and (C) to readd Pd–H in the opposite sense, giving (B) and hence (C) or (A).

The existence of the reversible reaction, (A) ⇌ (B), was shown by the behavior of α-methylstyrene **(I-9)** under the same conditions. Here the products were **(I-8)** (57%) and **(I-10)** (43%). Interestingly, in this reaction β elimination of a benzylic hydrogen was not preferred over β elimination of a terminal methyl hydrogen to the same extent as in the phenylation of propylene.

Very similar results were also obtained in the carbomethoxylation of the methylstyrenes using MeOCOPdOAc [from MeOCOHgOAc and $Pd(OAc)_2$]

† Anti-Markovnikov addition is even more favored in the presence of chloride ion; acetonitrile as solvent tends to favor Markovnikov addition.

TABLE I-2

Products from the Phenylation of Methylstyrenes with PhPdOAc in Methanol[a]

Starting material	Products (%): Me(Ph)C=CH(Ph)	Ph(Me)C=CH(Ph)	CH_2=C(Ph)($PhCH_2$)	Ph(Ph)C=CH(Me)
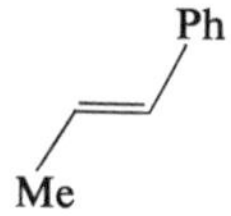 trans-MeCH=CHPh	98	Trace	0.5	Trace
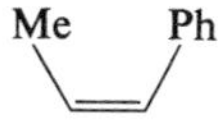 cis-MeCH=CHPh	22	65	10	3
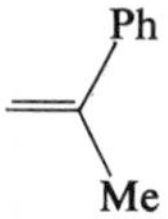 CH_2=C(Ph)Me	57	—	43	—

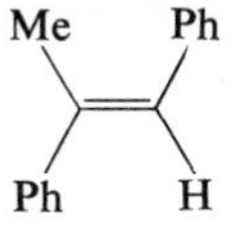

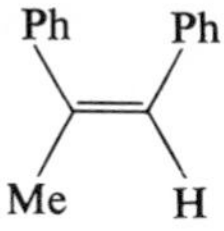

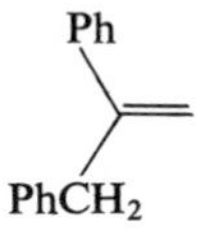

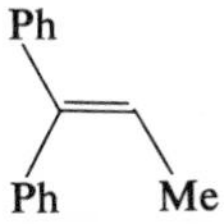

[a] After Heck.[25]

in acetonitrile. A mechanism involving cis addition and cis elimination was also proposed for this reaction.

Reactions very closely related to these are the Pd(II)-catalyzed arylations of olefins which can be carried out without the use of a separate organometallic arylating reagent.

$$C_6H_6 + RCH{=}CH_2 + Pd(OAc)_2 \longrightarrow (R)(H)C{=}C(H)(Ph)$$

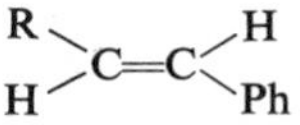

These reactions are further discussed below in Section D. The mechanisms suggested by Heck are of great utility in explaining a variety of other reactions of this type.

A reaction only somewhat tenuously linked mechanistically to the others described here, but of great potential practical importance is the hydrocyanation of olefins:

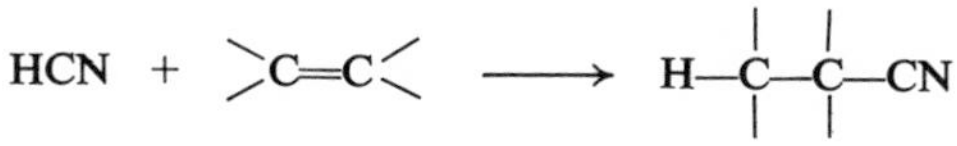

$$HCN + >C{=}C< \longrightarrow H{-}\overset{|}{\underset{|}{C}}{-}\overset{|}{\underset{|}{C}}{-}CN$$

SCHEME I-1

Both cobalt[26] and nickel[27] complexes have been used as catalysts, but Brown and Rick[28,29] have claimed that under specific conditions $[(PhO)_3P]_4Pd$ was an order of magnitude more effective. This catalyst alone, however, was poor, and the reaction was best carried out in the presence of approximately equivalent ratios of triphenylphosphite and HCN. This was accomplished by adding HCN slowly to the olefin, catalyst, and triphenylphosphite so that the quantity of unreacted HCN present in the reaction mixture was always less than that of triphenylphosphite. Experiments with norbornenes and norbornadiene showed that the system was very sensitive to steric effects. Norbornenes underwent addition so that the incoming CN group was always exo.

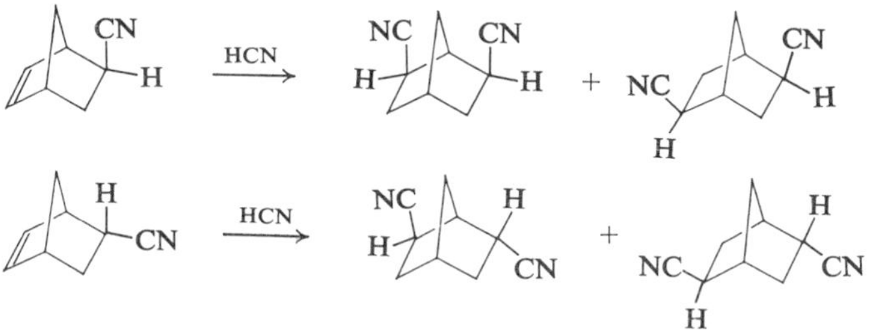

However, norbornadiene gave a small amount (16%) of the endo adduct, **(I-13)**. The difference was explained in terms of the vinylic hydrogens in

(I-13)

norbornadiene being sterically less active than the endo hydrogens in the norbornenes. Simpler olefins (ethylene, propylene) also underwent these reactions.

The reaction is visualized to proceed by a "classical" path involving oxidative addition of HCN to a Pd(0) complex, coordination of olefin, insertion of the olefin into the Pd–H bond to give a Pd–alkyl, and reductive elimination, aided by excess phosphite ligand, to regenerate the catalyst and to give the product $[L = (PhO)_3P]$:

$$L_4Pd \rightleftharpoons L_3Pd + L$$

$$L_3Pd + HCN \rightleftharpoons L_2Pd(H)CN + L$$

$$L_2Pd(H)CN + \text{olefin} \rightleftharpoons LPd(\text{olefin})(H)CN$$

$$LPd(\text{olefin})(H)CN + L \rightleftharpoons L_2Pd(\text{alkyl})CN$$

$$L_2Pd(\text{alkyl})CN + L \rightleftharpoons L_3Pd + \text{alkyl–CN}$$

In the absence of olefin, HCN reacted rapidly with the catalyst to give the catalytically inactive dicyano complex, $[(PhO)_3P]_2Pd(CN)_2$, presumably via a Pd(IV) dihydride intermediate.

$$L_2Pd(H)CN \xrightleftharpoons{HCN} L_2Pd^{IV}H_2(CN)_2 \longrightarrow L_2Pd(CN)_2 + H_2$$

It was thought that the excess triphenylphosphite was necessary to prevent the addition of a second mole of HCN.

The oxidative addition of HCN to Ir(I) has been reported by Singer and Wilkinson[30]; recently Drinkard *et al.*[31] and Schunn[32] have reported that some tetrahedral phosphine and phosphite nickel(0) complexes reacted with strong acids (HCl, CF_3COOH) to give 5-coordinate nickel hydrides. See also Volume I, Chapter II, Section D.

$$L_4Ni + HX \rightleftharpoons [L_4NiH]^+X^-$$
$$[L = (EtO)_3P, \tfrac{1}{2}(Ph_2PCH_2CH_2PPh_2)]$$

A five-coordinate structure, $L_3Pd(H)CN$, for the intermediate in the palladium-catalyzed hydrocyanation is quite plausible too.

Odaira *et al.*[33] have described a variant of this, using $Pd(CN)_2$ in a polar solvent, such as benzonitrile. Ethylene was converted to a mixture of acrylonitrile (51 %) and propionitrile (7 %) in this solvent. This reaction did not appear to be catalytic. Nonpolar solvents and triphenylphosphine repressed

$$C_2H_4 + Pd(CN)_2 \xrightarrow[PhCN]{55\ atm/150^\circ} CH_2{=}CHCN + EtCN$$

the reaction, while $K_2Pd(CN)_4$ was wholely inactive. The cyanation and hydrocyanation of ethylene and propylene with HCN in the presence of metallic palladium (on carbon) catalysts has also been described in two patents.[34,35]

$$C_2H_4 + HCN\ [+ O_2] \xrightarrow{315^\circ/Pd\text{–}C} CH_2{=}CHCN + EtCN$$

Other metals (on carbon) gave lower conversions.[35]

The formation of palladium metal and HCN in the reactions described by Odaira *et al.*[33] suggests that their reaction and those described in the patents are basically similar.

B. CARBONYLATION

The reactions considered next are those in which carbon monoxide is added to an olefin or similar substrate in various ways. There are some similarities between these reactions and the foregoing, particularly the cyanation reactions, as will emerge. The chief difference is that palladium carbonyls are rare and poorly characterized and have not, in general, been used as starting materials. There is, however, a considerable literature on carbonylation by other metal carbonyls, particularly of cobalt[36] and manganese.[37]

All these reactions can be interpreted in terms of the intermediate formation of an alkyl (or aryl) metal–carbonyl complex which undergoes a rearrangement, probably solvent- or ligand-assisted, to give an acylmetal complex by alkyl *migration* from the metal onto the carbon of the coordinated CO. The acyl is then cleaved from the metal in a number of ways to give aldehydes, acids, acid halides, esters, ketones, etc., depending on the conditions.

$$L_nMR(CO) \xrightarrow{L'} L_nL'MC(=O)R \rightarrow RCOOR',\ RCOCl,\ RCHO$$

Cossee[38] has discussed these and similar rearrangements theoretically and has come to the conclusion that they were all examples of "cis-migration rearrangements"; in other words the reaction involves migration of the alkyl group onto the carbon of coordinated CO rather than by insertion of CO into the alkyl-metal bond. This has been proved by the very elegant studies of Noack and Calderazzo on the carbonylation of methylmanganese pentacarbonyl and the reverse reaction.[37] Cossee suggested that the migrating ligand remained bonded to the metal throughout the reaction. However, the term "cis-insertion" (or even "insertion") is very commonly used to describe these reactions.

1. Alkyls and Aryls

Booth and Chatt[39, 40] reported the carbonylation of bis(triethylphosphine)-palladium and -platinum alkyl (and aryl) complexes **(I-14)** to the acyl (or benzoyl) complexes **(I-15)**. The palladium complex **(I-14a)** was carbonyl-

$$trans\text{-}(Et_3P)_2MRX + CO \rightleftharpoons trans\text{-}(Et_3P)_2M(COR)X$$

(I-14a) (M = Pd) **(I-15a)** (M = Pd)
(I-14b) (M = Pt) **(I-15b)** (M = Pt)

ated readily at atmospheric pressure and 20°, but the platinum analog **(I-14b)** needed pressure at 90°. The Pt–Me complex was carbonylated more readily than Pt–Ph. When X was alkyl (e.g., R = X = Me) rather than halogen, decomposition resulted; however, *cis*-dimethylbis(triethylphosphine)platinum did give some biacetyl (MeCOCOMe). The reactions leading to **(I-15)** were reversed on heating.

In this case the intermediate is likely to be a five-coordinate carbonyl $(Et_3P)_2M(CO)RX$; the relative ability of Pd and Pt to form five-coordinate complexes is then reflected in the more stringent conditions needed for carbonylation of the Pt complexes.

Medema *et al.*[11] have followed the carbonylation of the Cope complex, **(I-16)**, to **(I-17)**, by observing the carbonyl stretching bands in the infrared spectrum and suggested that the reaction proceeded as follows:

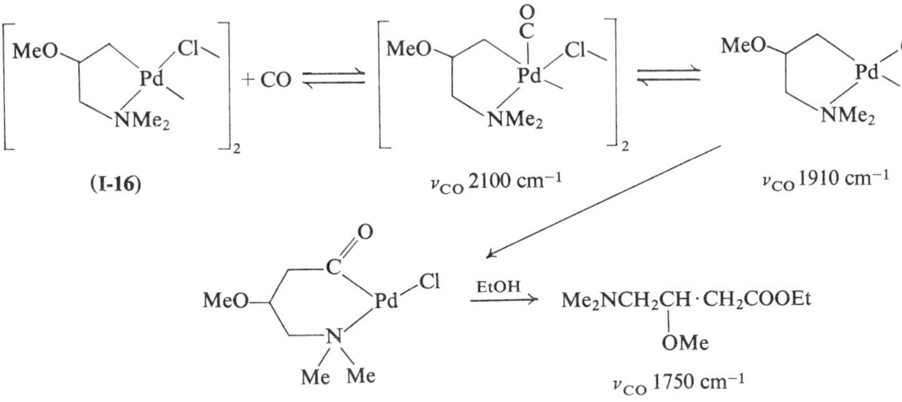

(I-16) + CO ⇌ ν_{CO} 2100 cm^{-1} ⇌ ν_{CO} 1910 cm^{-1} → $\xrightarrow{EtOH}$ $Me_2NCH_2CH(OMe)\cdot CH_2COOEt$ ν_{CO} 1750 cm^{-1}

(I-17)

The evidence is, however, not conclusive; the structural assignment of palladium carbonyls based on their infrared spectra alone is extremely dubious (Volume I, Chapter II, Section B,1).

Analogously, Takahashi and Tsuji[41] have reported the carbonylation (under much more drastic conditions) of a complex derived from azobenzene.

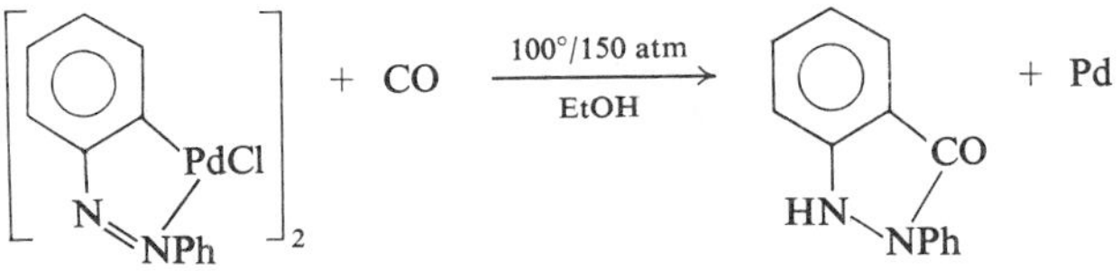

$[C_6H_4(PdCl)N{=}NPh]_2$ + CO $\xrightarrow[EtOH]{100°/150\ atm}$ $C_6H_4(CO)(HN)NPh$ + Pd

Stille and Haines[42] have reported the carbonylation of the *exo*-methoxynorbornenyl complex **(I-18)** in methanol to give 3-*endo*-carbomethoxy-5-*exo*-methoxynortricyclene. The same product was formed from norbornadienepalladium chloride and CO in methanol, but in benzene an uncharacterized intermediate [probably **(I-19)**, see Section B,4] was obtained,[43, 44] which with methanol gave 3-*endo*-carbomethoxy-5-*exo*-chloronortricyclene. The rearrangement to a nortricyclene is reminiscent of

PdCl OMe (I-18) $\xrightarrow{CO/MeOH}$ COOMe OMe $\xleftarrow{CO/MeOH}$ $PdCl_2$ $\xrightarrow{CO/C_6H_6}$ COCl Cl (I-19) $\xrightarrow{MeOH}$ COOMe Cl

the reaction of (**I-18**) with diphos[5] (see p. 5).

Arylmercuric halides reacted with palladium halides to give the biaryl or aryl halide (presumably via RPdX, see later); as already mentioned, this reaction was largely suppressed in the presence of olefins. Henry[45] and Heck[20] have observed that carbon monoxide also suppressed this reaction. Henry obtained benzoyl chloride, from phenylmercuric chloride and palladium chloride in acetonitrile (CO at 1 atm, 20°), esters in the presence of alcohols, or acids in the presence of acetic acid. The reactions could be made catalytic if $CuCl_2$ or $Hg(OAc)_2$ was added to reoxidize the palladium. Some ketones were also formed. Similar results were reported by Heck, who found that Rh(III) and Rh(I) complexes were even more active.

$$PhHgCl + PdCl_2 \longrightarrow \text{“PhPdCl”} \xrightarrow{CO} \text{“PhPdCOCl”} \longrightarrow Ph_2CO$$

$$\text{“PhPdCOCl”} \xrightarrow{HOAc} PhCOOH \qquad \text{“PhPdCOCl”} \xrightarrow{ROH} PhCOOR \qquad \text{“PhPdCOCl”} \longrightarrow PhCOCl$$

As in the reactions described earlier by Heck, other arylating agents (Ph_2Hg, Ph_4Sn, Ph_4Pb) could also be used.[45]

Garves[46] was able to generate arylpalladium species in situ by decomposition of arylsulfinates in the presence of $PdCl_2$. Esters and ketones were obtained when CO was added to the system.

$$ArSO_2Na + PdCl_4^{2-} + CO \xrightarrow{MeOH} ArCOOMe + Ar_2CO + SO_2$$

(Ar = phenyl, *p*-tolyl).

A patent has described the direct carbonylation of aryl halides[47]; benzoyl chloride was obtained in 80% yield from chlorobenzene in the presence of palladium chloride (CO, 160°/80 atm). Chlorides and bromides reacted more easily than iodides or fluorides and a cocatalyst (Ph_3P or PhCN) was desirable.

$$PhCl + CO \xrightarrow{PdCl_2} PhCOCl$$

2. Allyls

Since π-allylic complexes can be regarded as reacting via their σ-allyl forms, particularly in the carbonylation reactions, they are considered next.

These reactions were first reported by Tsuji and his collaborators[48–53]; they were also studied by Long and Whitfield[54] and Medema *et al.*[11] Their general form is:

$$[R'(\pi\text{-allyl})(R)PdCl]_2 + CO \xrightarrow{C_6H_6} RCH{=}CR'CH_2COCl$$
$$\xrightarrow{EtOH} RCH{=}CR'CH_2COOEt$$

For these reactions, pressure (50–200 atm) and moderate temperatures were needed. The CO always attacked the *least* substituted carbon atom [as did allene, but in contrast to butadiene; see this chapter, Section A; Volume I, Chapter V, Section F,12], and polar solvents increased the rate (DMF > C_6H_6 > CCl_4).

Occasionally collapse of the complex rather than, or together with, carbonylation, was observed, e.g.,[13,49]

$$(\pi\text{-allyl})Pd(acac) \xrightarrow{CO} Pd + CH_2{=}CHCH_2CH(COMe)_2$$

$$[(\pi\text{-}CH_2ClC_3H_4)PdCl]_2 \xrightarrow{CO/C_6H_6} ClCH_2CH{=}CHCH_2Cl + ClCH_2CH{=}CHCH_2COCl$$
$$\xrightarrow{CO/EtOH} ClCH_2CH{=}CHCH_2Cl + CH_2{=}CH(CH_2)_2COOEt$$

In the latter reaction a reductive carbonylation and isomerization, as well as esterification, was observed. The reactions of other chloroallylic complexes have been mentioned by Tsuji and Susuki.[52]

By analogy to those described above and to the reaction of π-allylic complexes with other nucleophiles (Ph_3P), these reactions can be rationalized as:

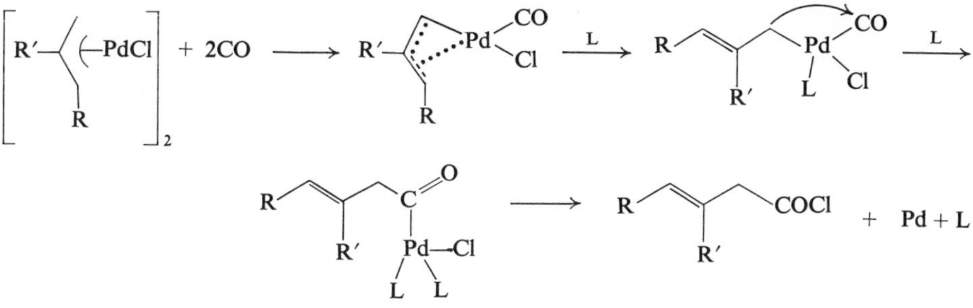

(L = chloride or solvent)

It is not clear whether the esters are formed from the acid chlorides, or directly, by nucleophilic attack of OEt^- on the Pd–acyl intermediate. It is also possible that a Pd–alkoxy species, in equilibrium with acyl–PdCl intermediates may be involved.

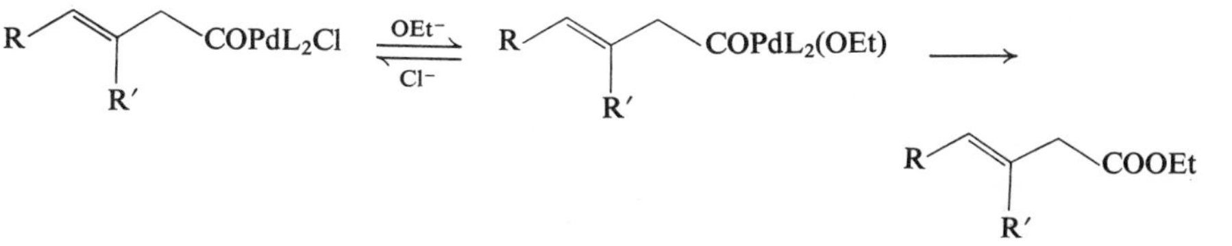

This reaction is important since it can be run catalytically to give the vinylacetic esters from allylic halides, carbon monoxide, and an alcohol.

$$RCH{=}CR'CH_2Cl + CO + R''OH \rightarrow RCH{=}CR'CH_2COOR'' + HCl$$

It is also catalyzed by nickel carbonyl,[55] but yields are much lower, owing to coupling reactions.[56]

A number of palladium complexes have been tested as catalysts. Medema *et al.*[11] found an order of reactivity: $[\pi\text{-}C_3H_5PdCl]_2 > (PhCN)_2PdCl_2 > PdCl_2 \gg Pd(\text{metal})$; other workers agree,[48] but a number of patents have used $PdCl_2$ as catalyst.[57–59] The rates for different allylic halides were allyl > 1-methylallyl ≈ 3-methylallyl ≫ cycloocten-3-yl,[11] and the reaction was found to be first-order with respect to both $[\text{allPdCl}]_2$ and allylic halide. Medema *et al.* also observed an induction period when Pd(metal), $PdCl_2$, or $(PhCN)_2PdCl_2$ were used as catalysts. They ascribed this to the necessity to form the active catalyst, the π-allylic complex, first. They also observed that a further rate enhancement was obtained using the allylic complex together with one mole of triphenylphosphine. More phosphine reduced the rate again (to zero, for $4PPh_3:Pd$). Tributylphosphine and $SnCl_2$ retarded the reaction. Only the π-allylic *halide* complexes were active, and the suggestion has been made that the reactive intermediates are chlorine-bridged dimers.[11]† The reaction

$$RCH{=}CR'CH_2X + CO \rightarrow RCH{=}CHR'CH_2COX$$

also proceeded for allyl ethers, alcohols, and acetates[50]; and Medema *et al.* reported relative rates as $X = Cl\ (1) > OMe\ (0.4) \approx OH \gg Br\ (0.08) > OAc\ (0.05)$.

Dent *et al.*[56] noted that alkyl and benzyl halides did not undergo this reaction, and that complications could arise from the reaction of allylic halides with CO in the presence of alcohols. For example, allyl chloride was carbonylated in methanol to give water and seven products, three of which were unidentified. The through reaction to the ester is, therefore, best carried out in two stages, formation of the acid chloride, followed by esterification.

$$CH_2{=}CHCH_2Cl + CO + MeOH \xrightarrow{[C_3H_5PdCl]_2}$$

$$\underset{(7\text{–}11\%)}{Me_2CHCOOMe} + \underset{(13\text{–}27\%)}{CH_2{=}CHCH_2COOMe} + \underset{(4\text{–}8\%)}{MeCH{=}CHCOOMe} + \underset{(7\%)}{MeCHClCH_2COOMe}$$

The formation of these by-products has been ascribed by Tsuji *et al.*[50] to the presence of HCl in the reaction, since they are absent in reactions in which HCl is not produced. The saturated esters may arise from propylene, formed by hydrogenolysis of the allyl chloride. This suggests that the vinylacetic esters are not formed by simple displacement on the acid chloride. Evidence that these catalytic reactions proceeded via π-allylic complexes was reported by Dent *et al.*[56] who found that 1-chloro-2-butene and 3-chloro-1-butene gave the same product, 3-pentenoyl chloride.

$$\left.\begin{array}{c} MeCH{=}CHCH_2Cl \\ CH_2{=}CHCH(Me)Cl \end{array}\right\} \longrightarrow \left[\pi\text{-}(C_3H_4Me)PdCl\right] \xrightarrow{CO} MeCH{=}CHCH_2COCl$$

Tsuji *et al.*[50] have also noted that allyl chloride reacted stoichiometrically with CO and $PdCl_2$ in benzene at 20° to give 3,4-dichlorobutyroyl chloride; they suggested that this occurred via a π olefin rather than a π-allylic complex, the formation of which required a higher temperature. At high conversions of

† Further studies by Volger *et al.*[60] led to the suggestion that a *tri*nuclear complex was involved. The only difference to the mechanism suggested above is that the chloride in the intermediates [e.g. (all)Pd(CO)(L)Cl] is now bonded to another $[(all)PdCl]_2$ unit and remains so throughout the reaction cycle.

allyl chloride, the metal was recovered as $[Pd_2Cl(CO)_2]_n$.[11] The carbonylation of some π-allylic complexes derived from 1,3-dienes is discussed in Section B,5.

3. Olefins

Four papers have discussed the reaction of (olefin)$PdCl_2$ complexes with CO in benzene.[11,44,61-63] The reaction again occurred easily and the products were the 3-chloroalkanoyl chlorides.

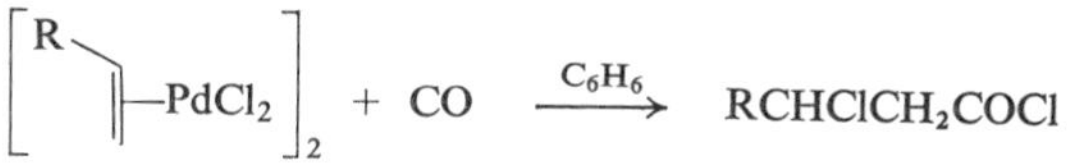

$$[R\text{-}CH{=}CH_2 \cdot PdCl_2]_2 + CO \xrightarrow{C_6H_6} RCHClCH_2COCl$$

Yields reported were low (C_2H_4 gave 41%, the others less); and CO again attacked at the least hindered site in the olefin. However, it is not clear just how much olefin complex was present in the reactions since they were carried out by carbonylating *in situ* a mixture of olefin and $PdCl_2$ in benzene. This solvent is not the best medium for formation of olefin complexes. Medema *et al.*[11] suggested the following reaction path:

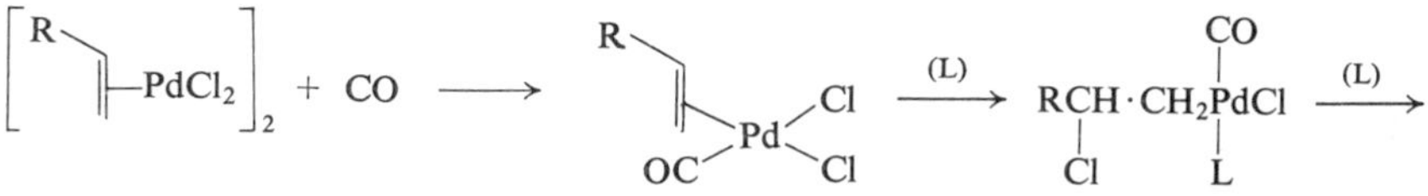

$$[R\text{-}CH{=}CH_2 \cdot PdCl_2]_2 + CO \longrightarrow (R\text{-}CH{=}CH_2)(OC)PdCl_2 \xrightarrow{(L)} RCH(Cl)\cdot CH_2Pd(CO)(L)Cl \xrightarrow{(L)}$$

$$RCHClCH_2COPdCl(L)_2 \longrightarrow RCHClCH_2COCl + Pd$$

$$(L = Cl^- \text{ or solvent})$$

Here again solvent effects appear to be very important. For example, a Shell patent[57] describes the carbonylation of ethylene in CCl_4 to give a mixture of three acids, after hydrolysis,

$$C_2H_4 + CO \xrightarrow[125\ atm/125^\circ]{PdCl_2/CCl_4} \underset{5}{CH_2{=}CHCOOH} + \underset{18}{ClCH_2CH_2COOH} + \underset{5}{Cl_3CCH_2CH_2COOH}$$

(5 : 18 : 5)

and a patent by Kutepow *et al.* describes the *di*carbonylation of ethylene in the presence of HCl,[64]

$$C_2H_4 + CO + HCl \xrightarrow[80^\circ/200\ atm]{C_6H_6} ClCOCH_2CH_2COCl + ClCH_2CH_2COCl$$

The most exhaustive study of these reactions in alcoholic solution is that of Bittler *et al.*[65-68] These authors tested a number of catalysts for the direct formation of esters from olefins. Again, nickel compounds (halides or carbonyl) had serious disadvantages, largely owing to side-reactions. Reactivity was shown by $(Ph_3P)_4Pd$, but only in the presence of HCl; better catalysts were $(Ph_3P)_2PdCl_2$ and $(Ph_3P)PdCl_2L$, where L was a weakly bound ligand such as an amine (piperidine, benzylamine). In this case reaction occurred below 100° in the presence of CO, an alcohol, and HCl. Other catalysts were briefly mentioned; substitution of triphenylphosphine by tri-*n*-butylphosphine decreased the rate, while for $(Ph_3P)_2PdX_2$ the effectiveness varied in the order $X = Cl > Br > I \sim SO_4$, NO_3, OAc. Unconjugated polyolefins underwent ethoxycarbonylation in a stepwise manner, higher temperatures favoring disubstitution.[66,67,69–71] No isomerization was observed.

Acids have also been obtained, and butadiene gave 3-pentenoic acid.[68] One remarkable feature is that in these reactions, in contrast to those carried out in benzene, attack "by CO" is now preferentially at the more heavily substituted carbon.[65,66] The reason for this is not clear.

$$CH_2{=}CHMe \xrightarrow[200\ atm/95^\circ]{CO/HCl/MeOH/(Ph_3P)_4Pd} \underset{(7}{Me(CH_2)_2COOH} + \underset{:\ 140\ :}{Me_2CHCOOMe} + \underset{82)}{Me(CH_2)_2COOMe}$$

Similar results were also reported by Tsuji *et al.*[72]

$$PhCH{=}CH_2 \xrightarrow[300\ atm/80^\circ]{CO/PdCl_2/EtOH/HCl} \underset{(2}{PhCH(Me)COOEt} + \underset{:\ \ 1)}{PhCH_2CH_2COOEt}$$

These authors also noted that a side reaction, the formation of the alkyl halide, became more important at higher HCl concentrations and for higher molecular weight olefins. Further studies are badly needed to clear up these very interesting effects.

An ICI patent[73] reported the production of α,β-, β,γ-, and γ,δ-homoangelica lactones by reaction of ethylene and CO at 2000 atm and 150° in acetonitrile in the presence of $PdBr_2$. Tsuji *et al.*[74] have reported that $PdCl_2$ (or Pd metal) will also catalyze the hydroformylation of ethylene (with H_2 and CO) to propionaldehyde but only in low yield; but see p. 74.

Fenton of Union Oil Co. has disclosed some details of a proposed process for the manufacture of acrylic acid by oxidative carbonylation of ethylene. The optimum conditions obtained at 127° in a solvent composed of pivalic and acetic acids, with some acetic anhydride present to remove water. The process was catalytic in the presence of $PdCl_2$–$CuCl_2$ and gave small amounts of β-acetoxypropionic acid as by-product.[75]

$$C_2H_4 + CO + [O] \rightarrow CH_2{=}CHCOOH$$

4. Unconjugated Dienes

Tsuji *et al.*[44,71] have reported the carbonylation of some diene complexes, but as these were not all carried out under the same conditions, direct analogies cannot be drawn.

Norbornadienepalladium chloride in benzene gave a nortricyclene, probably the isomer (**I-19**). 1,5-Cyclooctadienepalladium chloride in alcohol gave a mixture of saturated diesters and unsaturated monoester (see below.)

Under similar conditions the methoxycyclooctenyl complex (**I-20**) was carbonylated to methyl 3-methoxycyclooctenyl-7-carboxylate; 1,5-cyclooctadiene was also formed.[71]

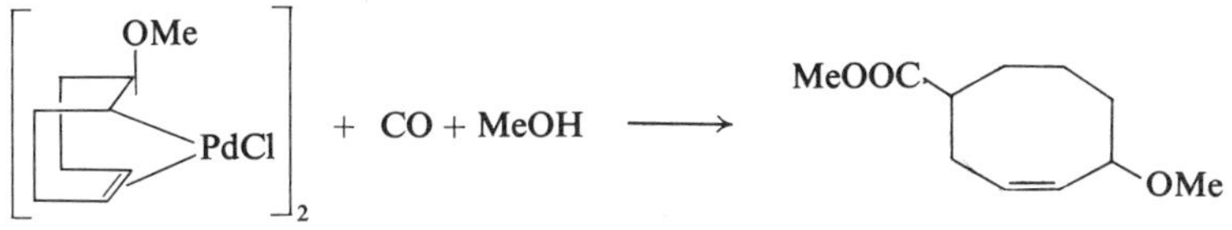

OMe PdCl]$_2$ + CO + MeOH ⟶ MeOOC ... OMe

(**I-20**)

Free 1,5-cyclooctadiene was carbonylated first to the unsaturated ester, and then to saturated diesters.[65–69,71,76] Catalysts which have been used for this include $PdCl_2/HCl/100°/13$ atm,[69,71] $(Bu_3P)_2PdI_2/THF/150°/1000$ atm,[77] $(Bu_3P)_2PdI_2/p\text{-}MeC_6H_4SO_3H/150°/250$ atm,[78] and $(Ph_3P)_2PdCl_2/HCl/120°/700$ atm[65]; presumably the phosphine complexes have a greater life as catalysts.

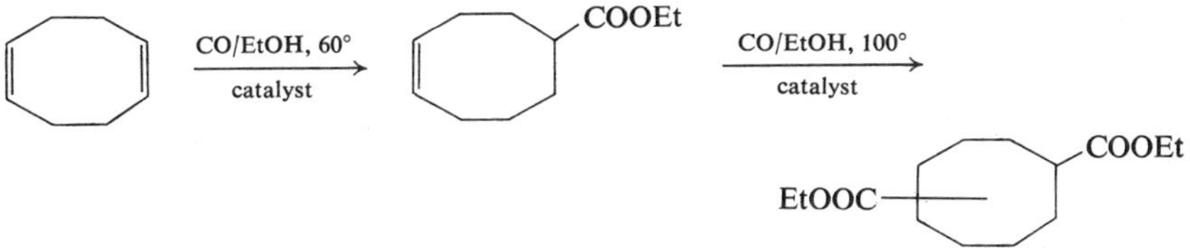

CO/EtOH, 60° / catalyst ⟶ COOEt ; CO/EtOH, 100° / catalyst ⟶ EtOOC— COOEt

In aprotic solvents (e.g., THF) Brewis and Hughes[77–79] found that 1,5-cyclooctadiene and unconjugated α,ω-dienes were carbonylated to substituted cyclic ketones in the presence of $(Bu_3P)_2PdI_2$.

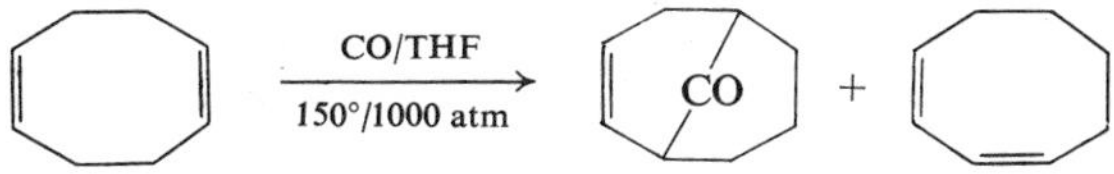

CO/THF / 150°/1000 atm ⟶ CO +

(**I-21**)

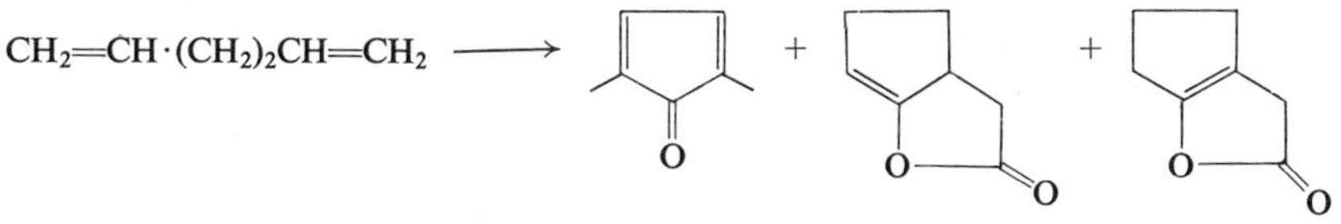

$CH_2{=}CH\cdot(CH_2)_2CH{=}CH_2$ ⟶ O + O O + O O

The formation of the unsaturated ketones, such as **(I-21)**, was explained by the following cyclic mechanism:

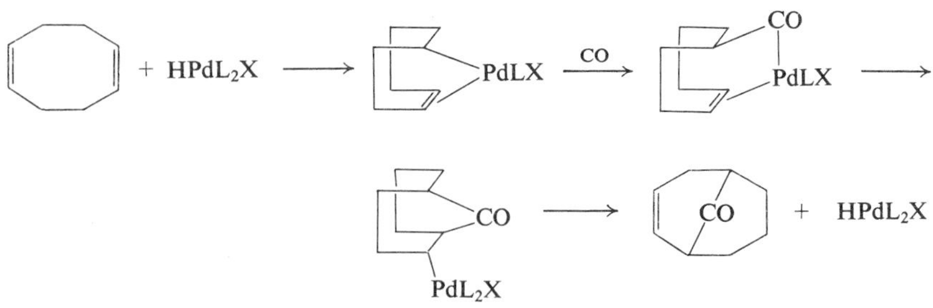

$$\text{C}_8\text{H}_{12} + \text{HPdL}_2\text{X} \longrightarrow \text{PdLX} \xrightarrow{\text{CO}} \text{CO}/\text{PdLX} \longrightarrow \text{CO}/\text{PdL}_2\text{X} \longrightarrow \text{CO} + \text{HPdL}_2\text{X}$$

The origin of $HPdL_2X$ ($L = Bu_3P$, $X = I$) is not clear, but sources of hydride are available in the system. The interesting step in the proposed mechanism is the insertion of the coordinated olefin into the Pd–C bond (or migration of the acyl group onto the coordinated olefin with concurrent formation of a σ bond). Models show that the geometry of the system is approximately correct for such a cis insertion to be highly favored here. In methanol, 1,5-hexadiene gave largely (45%) keto ester **(I-22)**; an analogous mechanism for its formation was proposed.[76,78]

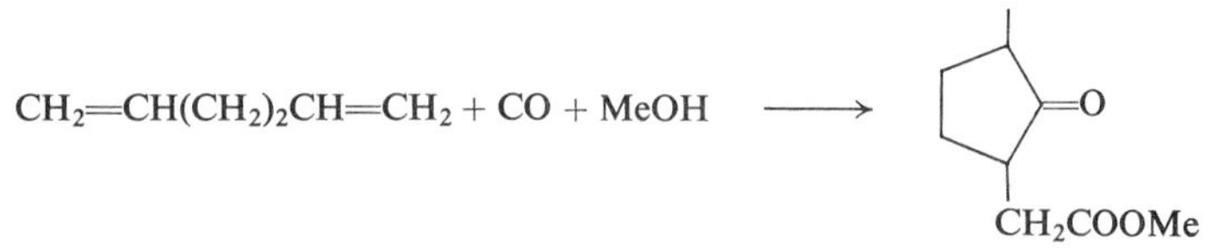

$$CH_2{=}CH(CH_2)_2CH{=}CH_2 + CO + MeOH \longrightarrow$$

CH_2COOMe

(I-22)

It is interesting that both here and in the cyclooctadiene reaction, the geometry of the Pd–*alkyl* intermediate proposed, in contrast to that of the Pd–acyl, is not favorable for a cis insertion, since the two sp^2 carbons and the Pd–C σ bond are not coplanar but perpendicular or nearly so. It therefore appears that the geometry of the intermediate complex plays a vital role in its further reactions. However, the validity of the proposed mechanism for both reactions is not clear. If it were really as simple as was suggested, then carbonylation of 1,5-cyclooctadiene in the presence of alcohol should also have given some of the bicyclic ketone **(I-21)**, since the formation of the keto ester **(I-22)** implies that insertion of coordinated olefin into Pd–acyl is fast compared with solvolysis of the Pd–acyl bond. Again, considerably more information than is available is necessary.

Interestingly, Brewis and Hughes[78,79] noted that the higher dienes such as 1,7-octadiene did not react in this way.

5. 1,2- and 1,3-Dienes

Both Brewis and Hughes[76,78,80] and Tsuji *et al.*[49,71,81] reported that 1,3-dienes gave β,γ-unsaturated esters under conditions similar to those discussed above. The yield of methyl 3-pentenoate from butadiene and CO in methanol at 1000 atm and 70° was shown to depend on the catalyst [$(Bu_3PPdI_2)_2$ (71%) > $(Bu_3P)_2PdBr_2$ (64%) > $(Bu_3P)_2PdI_2$ (60%) > Na_2PdI_4 (35–40%) > $(Bu_3P)_2PdCl_2$ (20%)].

$$CH_2{=}CHCH{=}CH_2 + CO + MeOH \rightarrow MeCH{=}CHCH_2COOMe$$

Other examples include:

$$\text{cyclopentadiene} + CO + MeOH \longrightarrow \text{cyclopentenyl}{-}COOMe$$

$$CH_2{=}CHC(Me){=}CH_2 + CO + MeOH \longrightarrow$$

$$\underset{(15\%)}{MeCH{=}CMeCH_2COOMe} + \underset{(38\%)}{Me_2C{=}CHCH_2COOMe} + \underset{(10\%)}{CH_2{=}CMe(CH_2)_2COOMe}$$

In contrast, Bordenca and Marsico[82] found that the $PdCl_2$- (or Pd) catalyzed carbonylation of piperylene in ethanol only gave ethyl 2-methyl-3-pentenoate. These carbonylation reactions were presumed to proceed via π-allylic complexes, e.g.,

$$CH_2{=}CHCH{=}CHMe \xrightarrow{PdCl_2/EtOH} \left[\pi\text{-allyl}(CH_2OEt){-}PdCl\right]_2 \xrightarrow{\text{hydrogenolysis}}$$

$$\left[\pi\text{-allyl}{-}PdCl\right]_2 \xrightarrow{CO} \left[\text{alkenyl}{-}COPdCl\right] \xrightarrow{EtOH} MeCH{=}CHCHMeCOOEt$$

Tsuji and Hosaka[81] have studied the reactions of the π-allylic complex (**I-23**), derived from isoprene, under various conditions. They obtained a multitude of products, the most important ones being those shown. In the

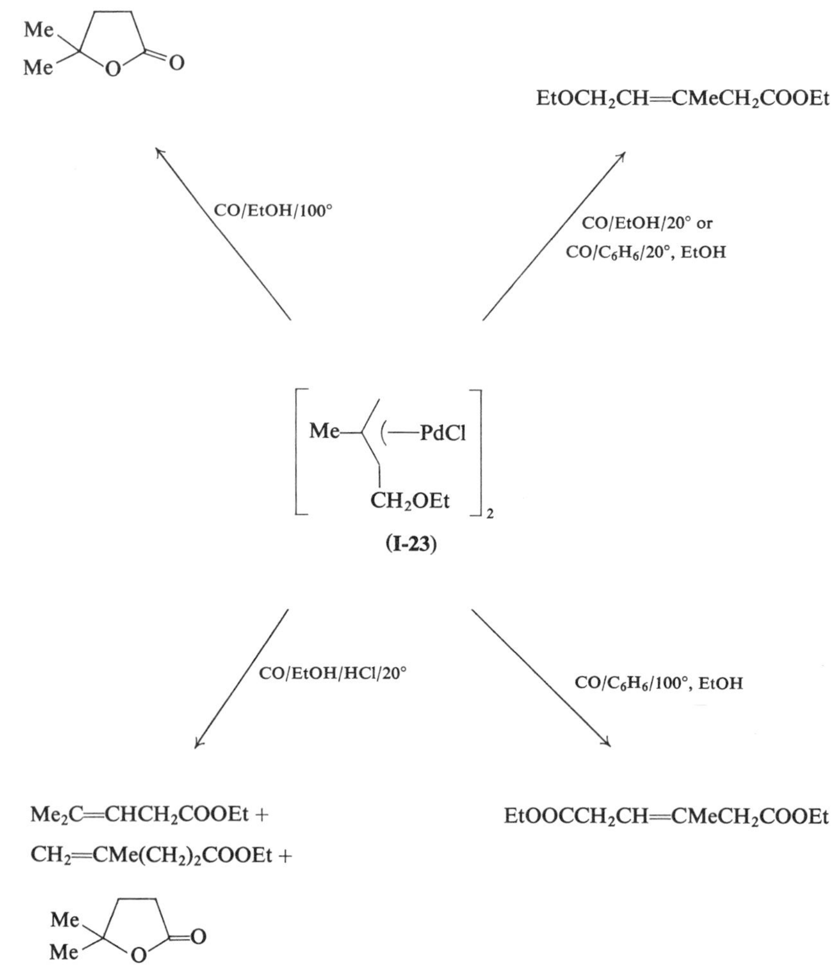

Me, Me (lactone, 5,5-dimethyl-γ-butyrolactone)

$EtOCH_2CH{=}CMeCH_2COOEt$

CO/EtOH/100°

CO/EtOH/20° or
$CO/C_6H_6/20°$, EtOH

$[Me{-}\langle(\text{—PdCl}) \; CH_2OEt]_2$

(I-23)

CO/EtOH/HCl/20°

$CO/C_6H_6/100°$, EtOH

$Me_2C{=}CHCH_2COOEt$ +
$CH_2{=}CMe(CH_2)_2COOEt$ +
Me, Me (lactone)

$EtOOCCH_2CH{=}CMeCH_2COOEt$

presence of HCl and in ethanol at 100° hydrogenolysis (of the ether) was important, and two of the products of the direct reaction of isoprene were obtained. Tsuji and Hosaka[81] suggested that attack at the unsubstituted allylic carbon was favored at low temperatures and led to $EtOCH_2CH{=}CMeCH_2COOEt$, while attack at the $-CH_2{-}O$ carbon predominated at higher temperatures, or when carbonium ion formation became favored; this led to $EtOOCCH_2CH{=}CMeCH_2COOEt$. The lactone was assumed to be derived from unsaturated esters. However, a detailed discussion of these reactions seems premature.

Other catalysts for carbonylation of 1,3-dienes have been reported.[65,68] Medema *et al.*[10,11] have also mentioned the coupling and carbonylation of butadiene and allyl chloride [to give (**I-24**)], catalyzed by allylpalladium chloride dimer.

$$CH_2{=}CHCH{=}CH_2 + CH_2{=}CHCH_2Cl + CO \xrightarrow{C_6H_6/90°/50\ atm} CH_2{=}CH(CH_2)_2CH{=}CHCH_2COCl$$

(I-24)

Susuki and Tsuji[83] have studied the carbonylation of allene and some complexes derived from allene. Allene itself gave a low yield of diethyl methylenesuccinate [$EtOOCC({=}CH_2)CH_2COOEt$] in ethanol.

6. Acetylenes

Tsuji *et al.*[84,85] and a German patent[86] described the carbonylation of acetylene catalyzed by $PdCl_2$. Dicarboxylic esters predominated, largely maleates and fumarates, although succinates were also formed. Tsuji *et al.*[84] also observed formation of a small amount of *trans,trans*-dimethyl muconate (**I-25**). Since acetylenes very readily oligomerize in the presence of Pd(II) this is not surprising.

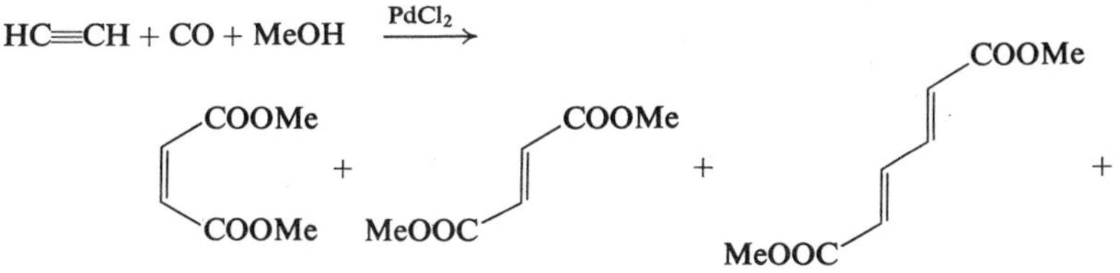

$$HC{\equiv}CH + CO + MeOH \xrightarrow{PdCl_2}$$

(cis-MeOOCCH=CHCOOMe) + (trans-MeOOCCH=CHCOOMe) + (trans,trans-MeOOCCH=CHCH=CHCOOMe, **(I-25)**) +

$$CH_2{=}CHCOOMe + MeOOCCH_2CH_2COOMe + MeCH_2COOMe$$

In their study of the carbonylation of diphenylacetylene, Tsuji and Nogi[87] showed that the main products were the lactone (**I-26**) (ca. 66%) and the diphenylmaleate (ca. 30%). No carbonylation occurred in benzene, HCl was necessary as a cocatalyst (at low concentrations of HCl a little diphenylacrylate was obtained) and the active catalytic species was probably *not* metallic Pd itself. The formation of the lactone was unexpected, and Tsuji and Nogi

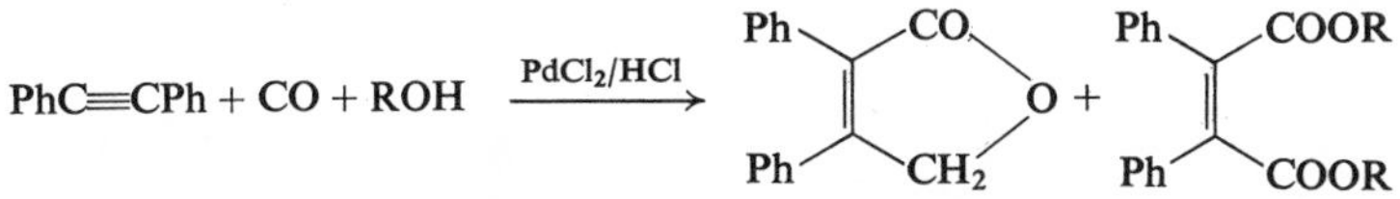

$$PhC{\equiv}CPh + CO + ROH \xrightarrow{PdCl_2/HCl}$$

(lactone: Ph(CO)C=C(CH_2)Ph ring closed through O, **(I-26)**) + (Ph)(COOR)C=C(Ph)(COOR)

suggested that it might be formed by hydrogenolysis of an intermediate analogous to the well-known cobalt complex[88] obtained from carbonylation of acetylene in the presence of $Co_2(CO)_8$.

$$(OC)_3Co—Co(CO)_3 \text{ bridged by } C(\text{lactone ring, } C{=}O,\ O) \text{ and } C{=}O$$

These authors have also studied the carbonylation of some monosubstituted acetylenes in acidic alcoholic media.[89–91] Some typical examples are given below.

$$HC{\equiv}CCH_2OH + CO \xrightarrow{Pd\text{–}C/MeOH}$$

$$CH_2{=}C(COOMe){-}CH_2COOMe + MeOOCCH_2CH(COOMe)CH_2COOMe + CH_2{=}C(COOMe)CH_2OMe$$

$$HC{\equiv}CCH_2Cl + CO \xrightarrow{PdCl_2/MeOH} CH_2{=}C(COOMe)CH_2COOMe + CH_2{=}C(Cl)CH_2COOMe$$

$$HC{\equiv}CCOOEt + CO \xrightarrow{Pd\text{–}C \text{ or } PdCl_2/EtOH/20^\circ} \textit{trans}\text{-}EtOOCCH{=}CHCOOEt +$$

$$(EtOOC)_2C{=}CHCOOEt + (EtOOC)_2CHCH_2COOEt +$$

$$(EtOOC)_2C{=}CH{-}CH{=}C(COOEt)_2$$

These reactions are very complex; dicarbonylation was usually favored, but products arising from mono- and tricarbonylation were found. In addition, reduction frequently accompanied these reactions and dimerization of the acetylene was not unknown. Benzenoid trimers were not obtained, but cyclic anhydrides and lactones were formed, particularly from reactions carried out in aprotic solvents.

7. Miscellaneous Carbonylation Reactions

Tsuji *et al.*[92] reported the carbonylation of cyclopropane.

$$\triangle + CO + PdCl_2 \xrightarrow{C_6H_6/90^\circ/13\ atm,\ MeOH}$$

$$Cl(CH_2)_3COOMe + MeCHClCH_2COOMe + EtCHClCOOMe + MeCH_2CH_2Ph$$

The authors suggested that the carbonylated products they obtained arose from propylene which was formed by isomerization of cyclopropane. However propylene normally only gives $MeCHClCH_2COOMe$, and the major product, $MeCH_2CHClCOOMe$, cannot arise by this route. Propylbenzene was only formed in the presence of CO.

Tsuji and Iwamoto[93] obtained ureas, oxamides, and formamides from primary amines and CO, catalyzed by $PdCl_2$ or Pd–carbon (see Chapter IV, Section B).

$$RNH_2 + CO \rightarrow RNHCONHR + RNHCOCONHR + RNHCHO$$

Tetraphenylcyclobutadienepalladium chloride reacted with CO to give dihydrotetraphenylcyclopentadienone.[94] With $Ni(CO)_4$ (and other metal carbonyls) in benzene, however, tetraphenylcyclopentadienone itself was very readily formed.[95] The difference in reaction is not understood.

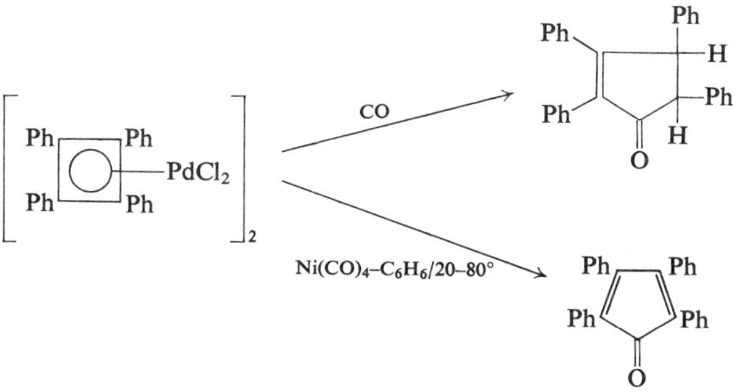

Ph Ph Ph H Ph H O
Ph Ph PdCl2 Ph Ph 2
CO
Ni(CO)4–C6H6/20–80°
Ph Ph Ph Ph O

Fenton and Steinwand in a patent[96] have described the formation of oxalate diesters from CO and an alcohol in the presence of $PdCl_2$, and Fe(III) or Cu(II).

$$2CO + 2ROH + PdCl_2 \rightarrow ROOCCOOR + Pd + 2HCl$$

Triethyl orthoformate was used as solvent and to remove any water formed.

Usami *et al.*[97] reported that palladium (especially when supported on carbon) catalyzed the carbonylation of cobalt(II) compounds to $Co_2(CO)_8$.

C. NONOXIDATIVE COUPLING AND OLIGOMERIZATION

A number of reactions are known in which organic compounds, usually olefins or dienes, couple (or oligomerize) in a palladium-catalyzed reaction which does not, overall, involve a reduction of the metal. These reactions are usually accompanied by isomerization (hydrogen transfer) reactions. Acetylenes, and π-allylic complexes also undergo coupling reactions easily; in many cases no hydrogen transfer occurs.

1. Dimerization of Monoolefins

The Pd(II)-catalyzed dimerization of ethylene to butenes and similar reactions have been known for a number of years. They were first noted, as side reactions in the Wacker process, by Smidt *et al.*[98] and were subsequently mentioned in a number of patents.[99–103] The dimerization occurred in anhydrous solvents in preference to the usual Pd(II) induced nucleophilic attack by water or other nucleophiles.

A number of studies have appeared of this reaction. The first, by Van Gemert and Wilkinson,[104] showed that the ethylenepalladium chloride complex, $[C_2H_4PdCl_2]_2$, decomposed slowly in solution (benzene or dioxane) to metal, butenes (1-, *cis*-2-, and *trans*-2-, in the ratio 10:25:65), polymer, and some chlorinated compounds (2-chlorobutane, 1,2-dichloroethane). Formation of the latter was favored by dioxane; benzene gave largely the butenes (63 %).

Kusunoki *et al.*[105] reported the catalytic dimerization of ethylene using $PdCl_2$ as catalyst in a variety of solvents. Acetic acid gave the highest yields (up to 1500 moles per $PdCl_2$); other solvents in order of decreasing effectiveness were: $AcOH > C_2H_4Cl_2 > C_2H_2Cl_4 > C_6H_6 > C_6H_5Cl > HO(CH_2CH_2O)_3H > ROAc > CHCl_3 > C_6H_{12} > EtOH$. Klein has mentioned that sulfolane in benzene is a good solvent.[103]

Ketley *et al.*[106] have carried out what is probably the most detailed study on this reaction; their results differ to some extent from those of Kusunoki *et al.* They found that under pressure ethylene reacted first with $PdCl_2$ to give $[C_2H_4PdCl_2]_2$, and that this reaction was markedly catalyzed by halo and nitro compounds as solvents, especially *t*-butyl or ethyl chloride. These solvents also strongly catalyzed the dimerization. In chloroform containing 0.03 *M* EtOH the complex was also formed, but it then slowly dissolved to give a red solution. This was catalytically highly active and dimerization of ethylene [to 1- (1 %), *trans*-2- (52 %), and *cis*-2-butene (47 %)] occurred. No dimerization occurred in the absence of ethanol (or water or other alcohols); ethylene glycol was an even more effective promoter. Ratios of alcohol:Pd > 1 deactivated the catalyst and metal was deposited. The authors postulated that the role of the alcohol here was to break the chlorine bridges in $[C_2H_4PdCl_2]_2$, in order to facilitate the formation of $(C_2H_4)_2PdCl_2$. A number of butene complexes were

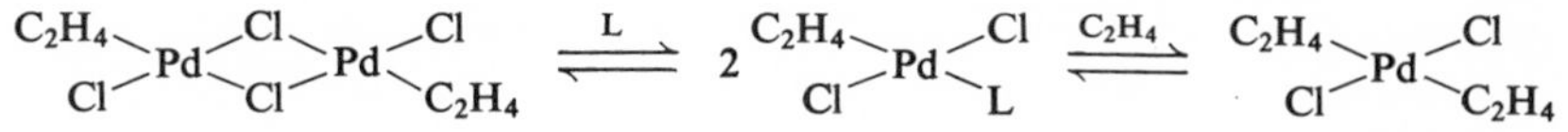

$$[C_2H_4(Cl)Pd(\mu\text{-}Cl)_2Pd(Cl)C_2H_4] \xrightleftharpoons{L} 2\ C_2H_4(Cl)Pd(Cl)L \xrightleftharpoons{C_2H_4} C_2H_4(Cl)Pd(Cl)C_2H_4$$

(I-27)

formed in these reactions and these are discussed in Volume I, Chapter II, Section E,6, but there is no real evidence to show how they are formed from (**I-27**). Ketley *et al.* suggested the intermediacy of a Pd(IV) hydrido complex, (**I-28**), but this is likely to be a rather high-energy intermediate. An alternative would be a catalytic cycle based on the presence of small amounts of solvated HPdCl in the system, which can possibly arise by a Wacker reaction of the olefin with traces of water or alcohol present.

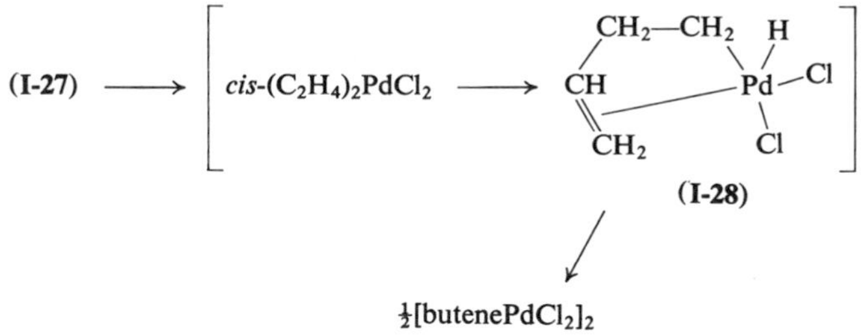

$$(\text{I-27}) \longrightarrow \left[cis\text{-}(C_2H_4)_2PdCl_2 \longrightarrow CH_2{=}CH\text{-}CH_2\text{-}CH_2\text{-}Pd(H)Cl_2 \right] \longrightarrow \tfrac{1}{2}[\text{butenePdCl}_2]_2$$

(I-28)

$$C_2H_4 + PdCl_2 + H_2O \rightarrow CH_3CHO + HCl + [HPdCl]$$

$$C_2H_4 + [HPdCl] \rightarrow [(C_2H_4)PdHCl \rightarrow C_2H_5PdCl]$$

$$[C_2H_5PdCl + C_2H_4 \rightarrow C_2H_5Pd(C_2H_4)Cl \rightarrow CH_3CH_2CH_2CH_2PdCl]$$

$$[CH_3CH_2CH_2CH_2PdCl \rightarrow (CH_3CH_2CH{=}CH_2)PdHCl] \rightarrow CH_3CH_2CH{=}CH_2 + [HPdCl]$$

$$\downarrow$$

$$[CH_3CH_2CH(CH_3)\cdot PdCl] \rightarrow CH_3CH{=}CHCH_3 + [HPdCl]$$

$$\text{butene} + PdCl_2 \rightarrow [\text{butenePdCl}_2]_2$$

The ethylene and butene complexes isolated are not then true intermediates.

Ketley *et al.*[106] found that addition of BH_4^- to the system deactivated it completely, but this observation does not appear to rule out the above mechanism.

Conti *et al.*[107] have reported the isolation of a form of $[\text{ethylenePdCl}_2]_2$ for which they suggested structure (**I-29**), and which in the solid (as well as in solution) gave butenes (largely *cis*-2-). The other isomers of $[C_2H_4PdCl_2]_2$ also gave this reaction, but induction periods were observed.

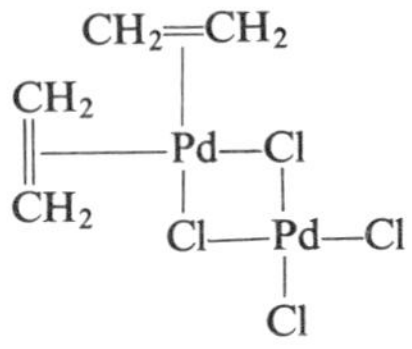

(I-29) 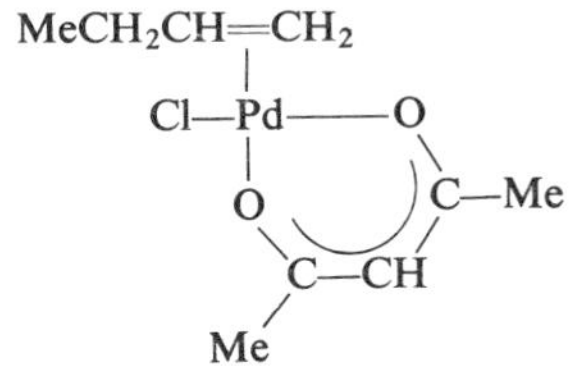 **(I-30)**

Kawamoto *et al.*[108] have obtained evidence (NMR at −45°) for the formation of a 1-butene complex **(I-30)** in the reaction of $[C_2H_4PdCl_2]_2$ with ethylene and acetylacetone in methylene chloride (see Volume I, Chapter III, Section E,6).

Butenes were also formed from ethylene in two curious reactions, described by Aguilo and Stautzenberger[109] and Crano *et al.*,[110] in which the main product was propylene (see p. 71).

A number of olefin dimerization reactions have been reported by Barlow *et al.*[111] using $(PhCN)_2PdCl_2$ or $PdCl_2$ at 100° either in dibutyl phthalate or without solvent. The reactions were quite slow, but the products in all cases were largely composed of linear olefins. For example, propylene gave hexenes and methylpentenes in a ratio of 11:1. A mixture of propylene and 1-butene gave mainly hexenes but also some coupled heptenes; similar reactions occurred with ethylene and styrene or methyl acrylate.

Kawamoto *et al.*[112] has shown the coupling of ethylene and styrene to be catalyzed by $[(PhCH{=}CH_2)PdCl_2]_2$ at 50°. The major product was *trans*-1-phenyl-1-butene, **(I-31)**, and the rate of the reaction depended on the solvent. In order of decreasing effectiveness these were: nitrobenzene, nitromethane > acetic acid, phenol > dioxane ≫ benzene > acetonitrile.

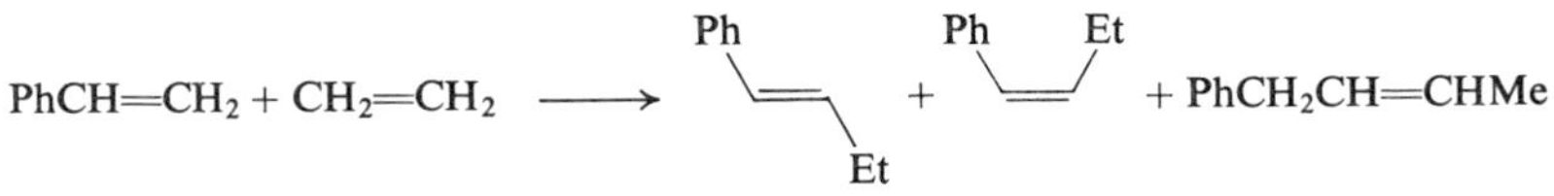

$$PhCH{=}CH_2 + CH_2{=}CH_2 \longrightarrow \text{(I-31)} + \text{cis-PhCH=CHEt} + PhCH_2CH{=}CHMe$$

(I-31)

The unusual $PdCl_2$-catalysed dimerization of 1-methylcyclopropene to give dimethyltricyclo[3.1.0.0^{2,4}]hexanes has been reported by Weigert *et al.*[115] 1,3,3-Trimethylcyclopropene reacted similarly but other cyclopropenes gave polymers.

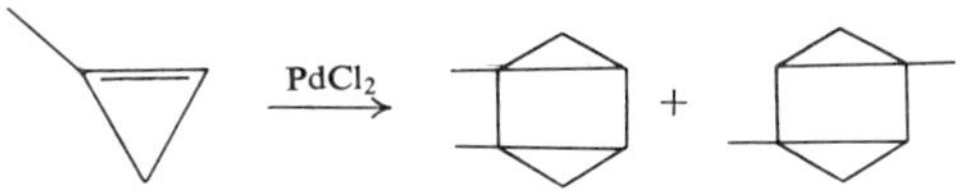

$\xrightarrow{PdCl_2}$ +

Dunne and McQuillin[116] have observed the coupling of some terpenes to be catalyzed by $(Ph_3P)_4Pd$ in chloroform.

$Pd(CN)_2$ polymerized ethylene and propylene.[100, 106, 117]

Klein[118] has reported the formation of ethyl *trans*-2-butenyl sulfone, **(I-32)**, together with some ethyl vinyl sulfone, from ethylene, SO_2, and palladium chloride in benzene suspension; about 6 moles per mole of $PdCl_2$ were

$$C_2H_4 + SO_2 \rightarrow \underset{\textbf{(I-32)}}{EtSO_2CH_2CH{=}CHMe} + EtSO_2CH{=}CH_2$$

formed. He noted that no reaction proceeded under rigorously anhydrous conditions and suggested that the intermediate might well be a palladium hydride, presumably arising again from reaction of a small amount of $[C_2H_4PdCl_2]_2$ with water. A patent[119] has claimed the formation of ethyl vinyl sulfone from ethylene, SO_2, and $PdCl_2$ under aqueous conditions. A possible mechanism for these reactions is:

$$C_2H_4 + PdCl_2 + H_2O \rightarrow CH_3CHO + HCl + [HPdCl]$$

$$[HPdCl] + C_2H_4 \rightarrow \left[C_2H_5PdCl \xrightarrow{SO_2} EtSO_2PdCl\right]$$

$$[EtSO_2PdCl + C_2H_4 \rightarrow EtSO_2CH_2CH_2PdCl] \rightleftharpoons EtSO_2CH{=}CH_2 + [HPdCl]$$

$$[EtSO_2CH_2CH_2PdCl + C_2H_4 \rightarrow EtSO_2(CH_2)_4PdCl]$$

$$\left[EtSO_2(CH_2)_4PdCl \rightarrow EtSO_2CH_2CH_2\underset{\substack{| \\ HPdCl}}{CH{=}CH_2} \rightleftharpoons EtSO_2CH_2CH_2\underset{\substack{| \\ PdCl}}{CH}CH_3\right]$$

$$[EtSO_2CH_2CH_2CHMePdCl] \rightarrow EtSO_2CH_2CH{=}CHMe + [HPdCl]$$

A feature common to many of these reaction mechanisms is an isomerization step in which a terminal PdCl is converted to an internal PdCl attached to a secondary carbon. This suggests that the bond between palladium and a secondary alkyl group may be thermodynamically more stable than that between palladium and a primary alkyl group under some circumstances.

Hafner *et al.*[120] have reported the formation of two five-membered cyclic alcohols in the reaction of allyl alcohol and palladium chloride. A mechanism has been proposed by these workers and Urry and Sullivan.[121]

$$CH_2{=}CHCH_2OH + PdCl_2 \longrightarrow$$ (3-methylene-tetrahydrofuran-2-methanol, CH_2, CH_2OH) + (3-methyl-2,5-dihydrofuran-2-methanol, CH_2OH) +

$$MeCH{=}CH_2 + MeCH_2CHO + CH_2{=}CHCHO + [C_3H_5PdCl]_2$$

2. Coupling of Monoolefins and Acetylenes or 1,3-Dienes

Related reactions involving coupling of α-olefins (particularly ethylene) and disubstituted acetylenes have been reported. Hosokawa *et al.*[122] showed that *t*-butyl(phenyl)acetylene reacted with $[C_2H_4PdCl_2]_2$ in benzene to give the π-allylic complex (**I-33a**). Mushak and Battiste[123] observed that diphenylacetylene reacted with ethylene and $(PhCN)_2PdCl_2$ in benzene to give *trans*,-*trans*-3,4-diphenyl-2,4-hexadiene (**I-34**); similar products were obtained from propylene, isobutene, 1-butene or styrene, and diphenylacetylene. A suggested mechanism for these reactions is given below and involves as first step the cis insertion of the acetylene into a Pd–Cl bond, followed by insertion of a coordinated olefin into a Pd–vinyl bond and a H migration. A similar mechanism has been proposed for the reaction of acetylenes with $PdCl_2$ (see below).

$$RC{\equiv}CR' + PdCl_2 \longrightarrow [RC{\equiv}CR'{-}PdCl_2] \longrightarrow RC(Cl){=}C(R'){-}PdCl \xrightarrow{CH_2{=}CH_2}$$

$$RC(Cl){=}C(R'){-}PdCl(CH_2{=}CH_2) \longrightarrow RC(Cl){=}C(R')CH_2CH_2{-}PdCl \longrightarrow$$

$$RC(Cl){=}C(R')CH{=}CH_2 \cdot PdHCl \longrightarrow [RC(Cl){=}C(R')CH(CH_3){-}PdCl] \longrightarrow [R(Cl)C{\cdots}C(R'){\cdots}CH(Me)\ PdCl]_2$$

(**I-33a**) (R = *t*-Bu, R′ = Ph)
(**I-33b**) (R = R′ = Ph)

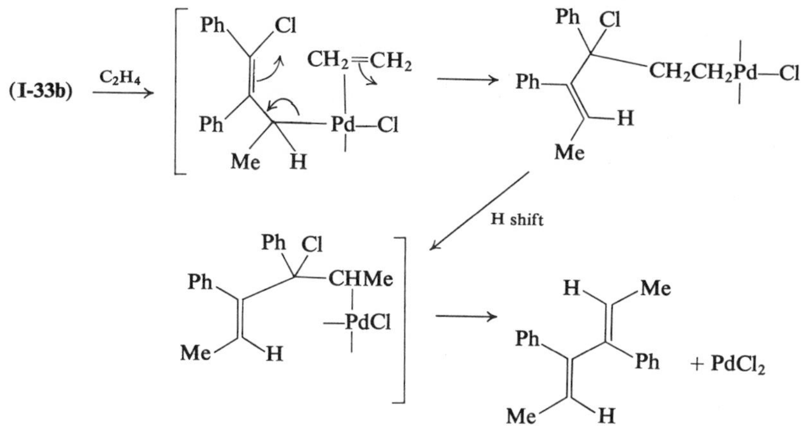

(I-33b) $\xrightarrow{C_2H_4}$ [Ph(Cl)C=C(Ph)–C(Me)(H)–PdCl · $CH_2{=}CH_2$] → Ph–C(=CH–Me...)C(Ph)(Cl)CH_2CH_2PdCl

H shift

[Ph(Cl)C(CHMe–PdCl)–C(Ph)=C(Me)H] → (H)(Me)C=C(Ph)–C(Ph)=C(Me)(H) + $PdCl_2$

(I-34)

In the formation of **(I-34)** no intermediate corresponding to **(I-33b)** has yet been isolated.

A brief account of the coupling of ethylene and butadiene to *trans*-1,4-hexadiene has been given by Schneider.[124] This reaction was catalyzed by

$$CH_2{=}CH_2 + CH_2{=}CHCH{=}CH_2 \rightarrow trans\text{-}CH_2{=}CHCH_2CH{=}CHMe$$

palladium chloride, particularly as $(Bu_3P)_2PdCl_2$, in the presence of Bu_2AlCl. No reaction occurred with tributylaluminum. Similar reactions were observed with other 1,3-dienes. They can be interpreted in terms of the primary formation of $(Bu_3P)_2Pd(Bu)Cl$ (Bu_3Al would form a dibutyl complex which would presumably decompose completely) which by β elimination gives the hydride $(Bu_3P)_2PdHCl$, the active catalytic species. This could, in the manner of many metal hydrides, add to 1,3-dienes to give allylic complexes, which could then add the ethylene to form the observed product and regenerate the catalyst.

$$L_2PdCl_2 + Bu_2AlCl \longrightarrow BuAlCl_2 + L_2PdBuCl$$

$$L_2PdBuCl \longrightarrow L_2PdHCl + MeCH_2CH{=}CH_2$$

$$L_2PdHCl + CH_2{=}CHCH{=}CH_2 \longrightarrow L_2PdCl(CH(Me)CH{=}CH_2)$$

$$L_2PdCl(CHMeCH{=}CH_2) \rightleftharpoons (\pi\text{-allyl, H, Me})PdClL_2 \rightleftharpoons L_2PdCl\text{–}CH_2CH{=}CHMe$$

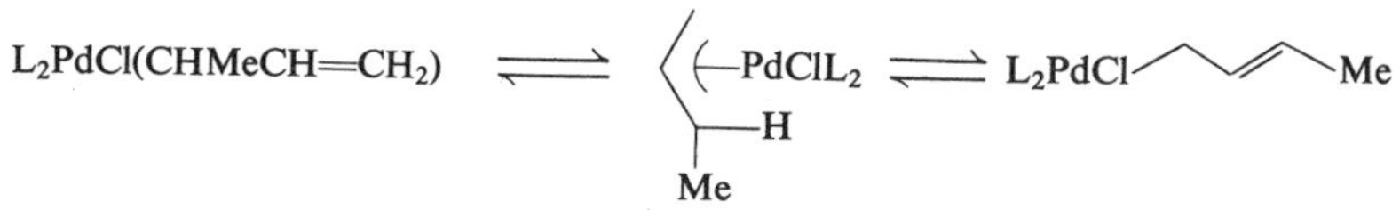

$$L_2PdCl\text{–}CH_2CH_2CH{=}CH\text{–}Me \xrightarrow{C_2H_4} L_2PdCl\text{–}CH_2CH_2CH_2CH_2CH{=}CH\text{–}Me$$

$$L_2PdCl\text{–}CH_2CH_2CH_2CH_2CH{=}CH\text{–}Me \longrightarrow L_2PdHCl + CH_2{=}CHCH_2CH{=}CHCH_3$$

The products obtained from isoprene and piperylene can be explained analogously; two modes of addition of Pd–H to the diene are then possible.[125] A somewhat analogous reaction scheme has been put forward by Cramer[126] for the Rh(III)-catalyzed synthesis of 1,4-hexadiene.

3. Coupling of Dienes

a. *Dimerization of a Cyclobutadiene*

Only one example of the coupling of a complexed diene has been reported. Maitlis and Stone[127] observed that tetraphenylcyclobutadienepalladium halides readily reacted with phosphines in benzene at 80° to give high yields of the cyclobutadiene dimer, octaphenylcyclooctatetraene[128] **(I-35)**. At 20°

$$[Ph_4C_4PdX_2]_2 + 4R_3P \xrightarrow{C_6H_6/80^\circ} Ph_8C_8 + 2(R_3P)_2PdX_2$$

(I-35)

the reaction mixture developed an intense green color, due to a paramagnetic species which Sandel and Freedman[129] identified as the halotetraphenylcyclobutenyl radical (see Volume I, Chapter IV, Section F,1,a). In the higher temperature reaction two of these radicals presumably combine and eliminate halogen to give octaphenylcyclooctatetraene. This mechanism implies a Pd(I) intermediate, and it would be interesting to obtain evidence for this.

Cookson and Jones[130, 131] have shown that when the reaction in benzene at 80° was carried out in the presence of methyl phenylpropiolate or cyclopentadiene, adducts were isolated.

$$[Ph_4C_4PdCl_2]_2 + PPh_3 \xrightarrow{PhC_2COOMe} C_6Ph_5COOMe + Ph_8C_8$$

$$[Ph_4C_4PdCl_2]_2 + PPh_3 \xrightarrow{C_5H_6} \text{tetraphenylindane adduct} + Ph_8C_8$$

b. *Coupling of Allene and a π-Allylic Complex*

π-Allylic complexes react with allene as shown.[11,132]

$$2CH_2{=}C{=}CH_2 + [R'C_3H_3(R){\leftarrow}PdCl]_2 \longrightarrow [RCH{=}CR'CH_2{-}C_3H_4{\leftarrow}PdCl]_2$$

$$CH_2{=}C{=}CH_2 + C_3H_5{\leftarrow}Pd(acac) \longrightarrow CH_2{=}CHCH_2{-}C_3H_4{\leftarrow}Pd(acac)$$

These reactions are discussed in Volume I, Chapter V, Section F,12. Coupling can also occur in the direct reaction of allene with palladium chloride to give **(I-37)**. Alternatively, depending on the solvent, **(I-36)** is formed (Volume I, Chapter V, Section B,2,b).

$$CH_2{=}C{=}CH_2 + PdCl_2 \longrightarrow \underset{\textbf{(I-36)}}{[Cl{-}C_3H_4{\leftarrow}PdCl]_2} + \underset{\textbf{(I-37)}}{[(CH_2{=})(ClCH_2)C{-}C_3H_4{\leftarrow}PdCl]_2}$$

Shier showed that allene reacted with Pd(II) complexes of weak ligands (OAc, NO_3, ClO_4) in acetic acid to give a wide variety of organic products.[7,133] The π-allylic acetate complexes were more reactive than the halides and Shier suggested this to be the reason why the reactions occurred catalytically here.

$$CH_2{=}C{=}CH_2 + Pd(OAc)_2 \xrightarrow{HOAc/65^\circ}$$
$$CH_2{=}CHCH_2OAc + CH_2{=}CMeCMe{=}CH_2$$
$$+ CH_2{=}CMeC({=}CH_2)CH_2OAc + AcOCH_2C({=}CH_2){\cdot}C({=}CH_2)CH_2OAc$$

The products can be rationalized to arise from an intermediate such as **(I-38)** analogous to **(I-37)** (Scheme I-2).

Shier also noted that in the presence of one mole of PPh_3 a polymer ($n = 10$–30) formulated as

$$\left[{-}CH_2{-}C({=}CH_2){-}C({=}CH_2){-}CH_2{-} \right]_n$$

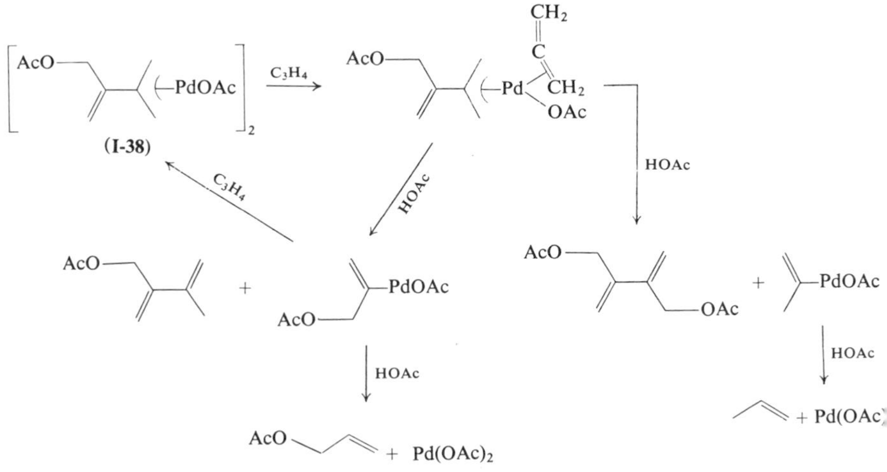

SCHEME I-2

was obtained.[7] A similar mechanism to the above was suggested, but it is not clear just why addition of PPh_3 caused the polymerization. In the presence of methylacetylene, allene reacted catalytically to give $MeC{\equiv}CCMe{=}CH_2$ and $MeC{\equiv}CC({=}CH_2)CH_2OAc$ as well as the products arising from allene alone. A palladium acetylide appears to be implicated as an intermediate.

c. *Coupling of 1,3-Dienes with Themselves and π-Allylic Complexes*

The reactions of π-allylic complexes with 1,3-dienes have already been discussed in Volume I, Chapter V, Section F,12. The reaction

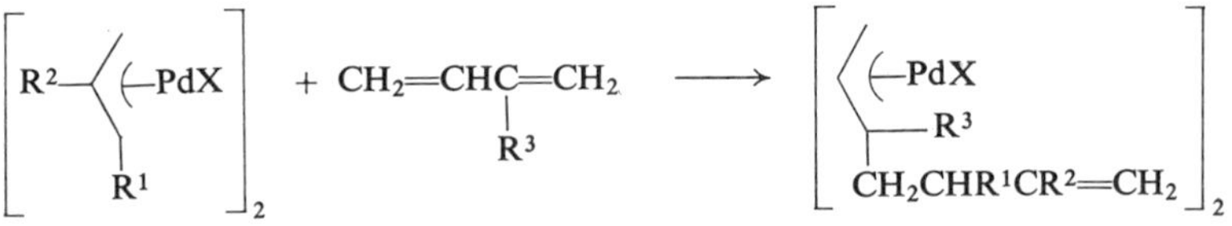

$$[R^2C_3H_3(R^1)PdX]_2 + CH_2{=}CHC(R^3){=}CH_2 \longrightarrow [C_3H_3(R^3)(CH_2CHR^1CR^2{=}CH_2)PdX]_2$$

was studied by a number of authors,[10–14] and proceeded particularly well for the acetates (X = OAc). In this case, however, a subsequent reaction involving a slow trimerization of butadiene to **(I-39)** and release of 1,3,7-heptatriene and 1,5-heptadiene, also occurred. Above 50° butadiene was catalytically trimer-

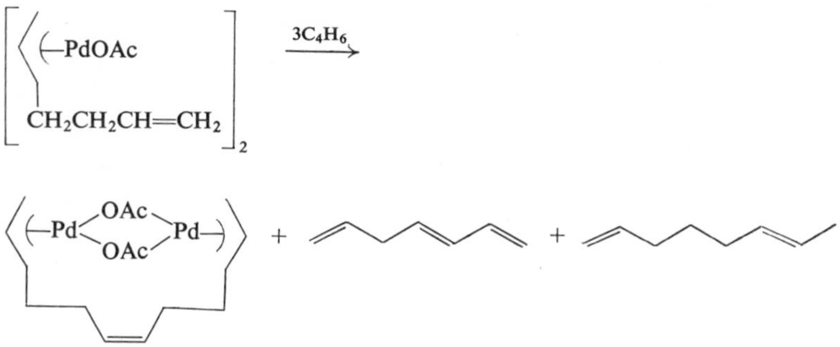

$$[(\text{C}_4\text{H}_6\text{PdOAc})(\text{CH}_2\text{CH}_2\text{CH}{=}\text{CH}_2)]_2 \xrightarrow{3\text{C}_4\text{H}_6}$$

(I-39)

ized by $[C_3H_5PdOAc]_2$ to 1,3,6,11-dodecatetraene; a small amount of dimer, 1,3,7-octatriene, was also formed.[11]

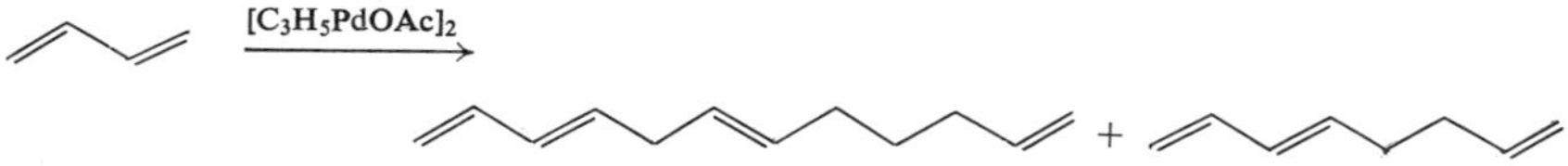

$$\xrightarrow{[\text{C}_3\text{H}_5\text{PdOAc}]_2}$$

Under different conditions 1,3,7-octatriene was the major product; catalysts for this included $(Ph_3P)_2Pd$(maleic anhydride) in benzene[134] or, better, acetone[135] or tetrahydrofuran.[134] A number of Pd(II) complexes including $PdCl_2$ and $[C_3H_5PdCl]_2$ in the presence of phenol and sodium phenate gave largely 1-phenoxy-*trans*-2,7-octadiene; on the addition of triphenylphosphine to the reaction mixture 1,3,7-octatriene was obtained.[136,137]

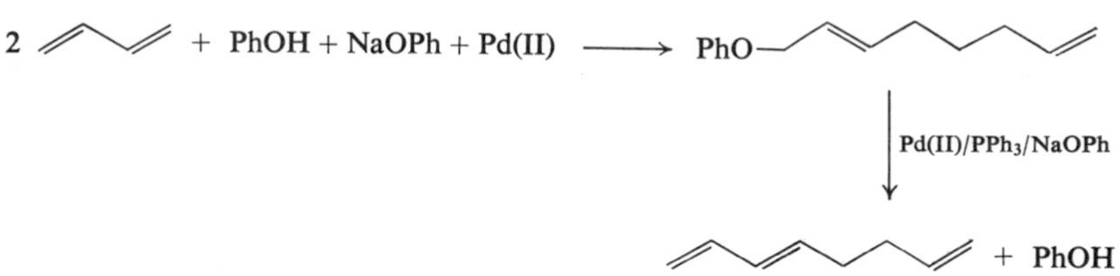

2 + PhOH + NaOPh + Pd(II) ⟶ PhO–

Pd(II)/PPh_3/NaOPh

+ PhOH

Hagihara and co-workers have shown that $(Ph_3P)_2Pd$(maleic anhydride) catalyzed the formation of 1-substituted *trans*-2,7-octadienes (**I-40**) from butadiene and organic compounds with active hydrogens (alcohols[134,135,138,139] phenols,[134] carboxylic acids,[135] or amines† [139]). Small amounts (5%) of the iso-

† Silanes (Me_3SiH) also reacted but gave 1-substituted 2,6-octadienes[140] (Chapter IV, Section D).

meric 3-substituted 1,7-octadienes were also obtained. With the higher alcohols the yields of **(I-40)** (X = RO) decreased until, for *n*-propanol, the major product (72%) was 1,3,7-octatriene. Using C_3H_7OD (70% d_1) monodeuterated 1,3,7-octatriene (60% d_1 and 40% d_0, D probably at C-6) was obtained, while CH_3OD gave 1-methoxy-6-deutero-2,7-octadiene.[138, 139] Other Pd(0) complexes, $(Ph_3P)_2Pd$(benzoquinone), $(Ph_3P)_2Pd$(dimethyl fumarate), and $(Ph_3P)_4.Pd$, were also active[138]; a communication by Kohnle *et al.*[141] noted that $(Ph_3P)_3Pt$, $(Ph_3P)_2PtCO_3$, or $(Ph_3P)_4Pd$, *all in the presence of CO_2*, gave good yields of the octatriene (see also Appendix).

A recent paper by Walker *et al.*[142] described the palladium acetylacetonate catalyzed reaction of butadiene and acetic acid to give largely 1-acetoxy-2,7-octadiene. A phosphine (PPh_3) or a phosphite $[EtC(CH_2O)_3P]$ was necessary as a co-catalyst, and the reaction was strongly promoted by equimolar amounts of certain tertiary amines, for example, $Me_2NCH_2CH_2OH$. In the presence of primary or secondary amines high yields of 2,7-octadienylamines were formed.

Bis(π-allyl)palladium has also been reported to catalyze the formation of *n*-dodecatetraene from butadiene.[143] In contrast, bis(π-allyl)platinum did not react and bis(π-allylic)nickel complexes gave 1,5,9-cyclododecatriene.

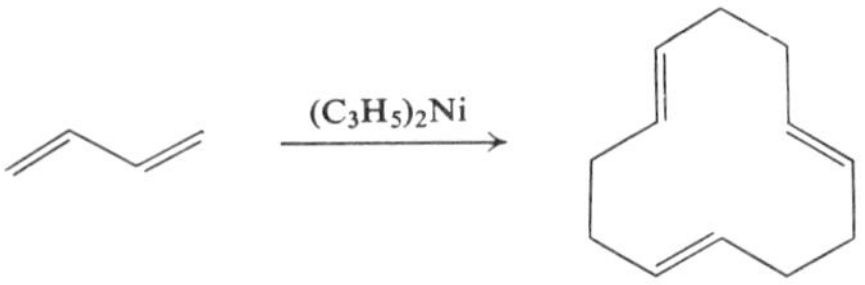

$$\xrightarrow{(C_3H_5)_2Ni}$$

Nickel(0) complexes in the presence of a tertiary phosphine or phosphite catalyzed the cyclodimerization of butadiene to divinylcyclobutane, 1,5-cyclooctadiene, and 4-vinylcyclohexene.[144, 145]

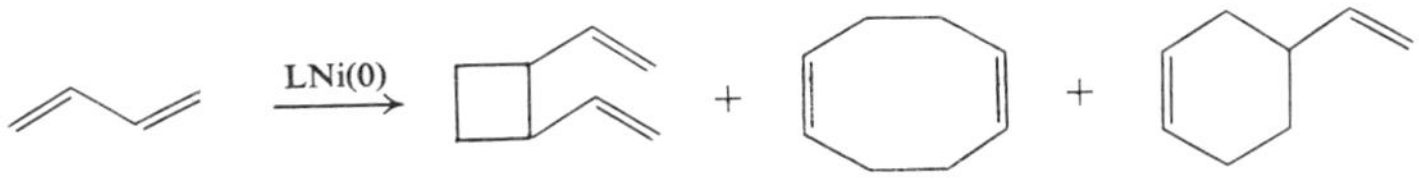

$$\xrightarrow{LNi(0)} \quad + \quad + $$

The intermediates in these cyclotri- and cyclodimerization reactions have been identified by Wilke and co-workers as acyclic bis(allylic)nickel complexes.[144]

Ni   L—Ni

Similar mononuclear intermediates have been suggested by Wilke,[143,145] Hagihara[138] and their co-workers to explain the Pd-catalyzed linear di- and trimerizations. In the formation of the hydrocarbons it was assumed that the size of the palladium atom precluded the cyclization, and that a rapid hydrogen transfer occurred instead. The 1-substituted 2,7-octadienes **(I-40)** were presumed to arise from an intermediate such as **(I-41)** (M = Pd) by 1,6-addition of X–H to the ligand, for example,

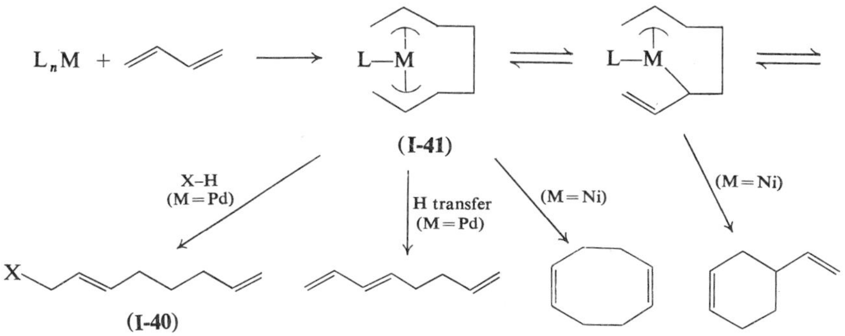

L_nM + [butadiene] ⟶ L—M **(I-41)** ⇌ L—M ⇌

X–H (M = Pd) → X-octadiene **(I-40)**; H transfer (M = Pd) → octatriene; (M = Ni) → cyclooctadiene; (M = Ni) → vinylcyclohexene

In all the reactions in which butadiene has been di- or trimerized using a π-allylic catalyst, the original π-allylic ligand comes off during the first cycle and subsequent cycles involve (formally) a reversible oxidation–reduction reaction between a butadiene–M(0) complex and a π-allylic–M(II) complex derived from butadiene [e.g., **(I-41)**]. The process by which this transformation occurs has been discussed.[146] Complexes such as **(I-39)** and

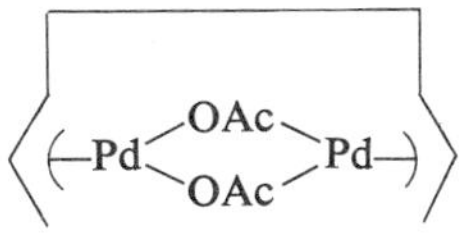

Pd(OAc)(OAc)Pd complex

in which two metal atoms participate, have also been reported and a modification of the above reaction scheme involving these species has been proposed by Medema *et al.*[10,11] to explain the formation of the linear dimers and trimers (see Appendix).

There is some evidence to suggest another route to the dimer products. The incorporation of deuterium at C-6 in both the 1,3,7-octatriene and the 1-methoxy-2,7-octadiene (see above) can also be explained by a stepwise mechanism involving the primary formation of a Pd(II)–hydride complex (active catalyst) from a Pd(0) complex by oxidative addition of X–H; for example,

$$L_nPd^0 + X{-}H \rightarrow L_2Pd(H)X \quad \text{(or, possibly, } [L_3PdH]^+X^-)$$

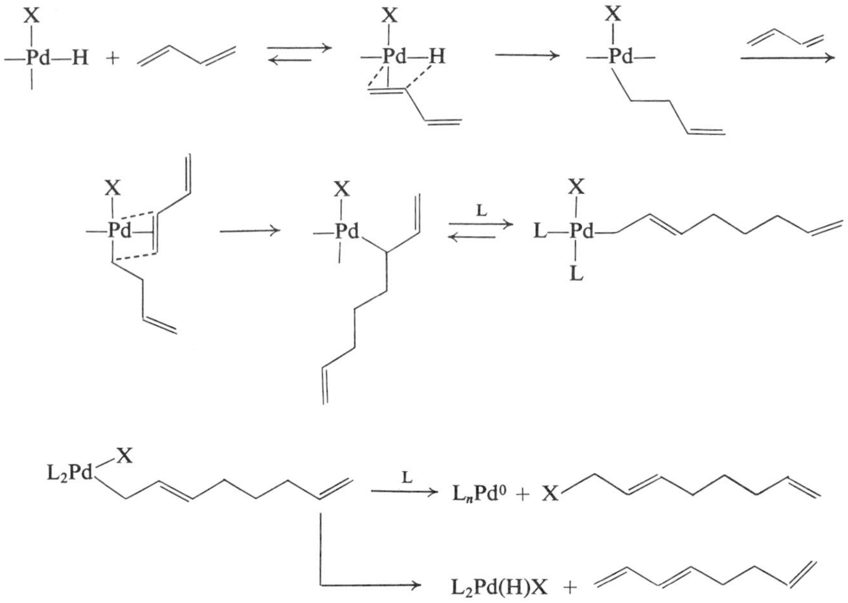

L

L

L

L_2Pd X

L_nPd^0 + X

$L_2Pd(H)X$ +

The somewhat unusual mode of addition to give the linear dimers is dictated by the presence of bulky ligands (e.g., Ph_3P) which preferentially stabilize primary over secondary alkylpalladium complexes as intermediates.

Manyik *et al.*[147] have reported that butadiene reacts with aldehydes in a $Pd(acac)_2$ catalyzed reaction,

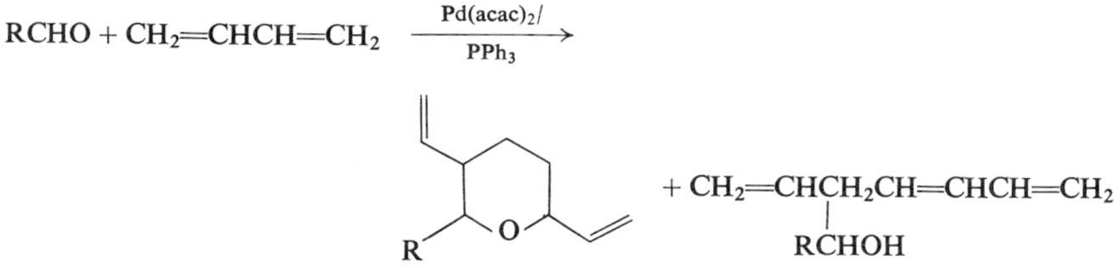

$$RCHO + CH_2{=}CHCH{=}CH_2 \xrightarrow[PPh_3]{Pd(acac)_2/}$$

$$+ \; CH_2{=}CHCH_2CH{=}CHCH{=}CH_2 \text{ (with RCHOH substituent)}$$

R O

The relative amounts of the two products for R = Me depended on the phosphine: palladium ratio; formaldehyde gave only the pyran (R = H).

Heimbach has also described the linear cotrimerization of butadiene and secondary amines to give dodecatrienylamines with a nickel(0) catalyst.[148] Under some conditions linear di-, tri-, or tetramers were obtained from butadiene and nickel compounds.[145, 148, 149]

The polymerization of butadiene by palladium(II) halides has been reported by Canale *et al.*[150, 151] 1,2-Polybutadienes and *trans*- and *cis*-1,4-polybutadienes were all produced; $PdBr_2$ gave 91% 1,2-, while PdI_2 gave 85% *trans*-1,4-, and $PdCl_2$ gave a mixture. Ligands such as R_3P, Me_2SO, and SCN^- deactivated the catalyst. Wilke has reported that butadiene can also be polymerized with π-allylnickel tetrabromoaluminate.[143]

It is also interesting to compare these dimerization reactions with the very specific cobalt-catalyzed dimerization of butadiene to 5-methyl-*trans*-1,3,6-heptatriene. The active catalyst appears to be a cobalt–allylic complex, isolated and fully characterized by X-ray analysis,[152] as **(I-42)**.

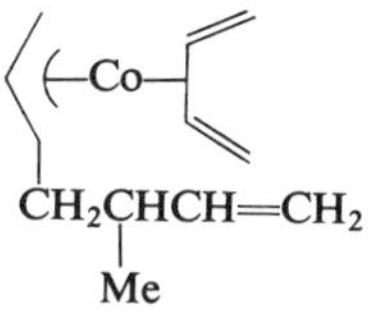

Co

$CH_2CHCH{=}CH_2$

Me

(I-42)

d. *Coupling and Oligomerization Reactions of Acetylenes*

Acetylenes react readily with many, if not most, transition metal complexes to give a bewildering variety of products and few mechanistic rationalizations have been possible as yet. Various aspects of this topic have been reviewed and discussed by a number of authors.[153–156]

Palladium(II) is no exception; in fact, it probably reacts with even greater facility than most other transition metals.

The reaction of aqueous $PdCl_2$ with acetylenes was first noted by Phillips[157] in 1894. Erdmann and Makowka[158] in 1904 used the precipitated complex formed from this reaction in acid at 20° as a method for estimating palladium in the presence of other platinum metals.† Makowka[161] in 1908 was the first to study the complexes formed, but the analyses were not always reproducible, and depended on the method by which the precipitate was purified. One sample, which analyzed for C_4H_5ClOPd gave butyric acid after heating with KOH and acidification; higher fatty acids may also have been present. Temkin *et al.*[162,163] were able to isolate a complex with the above analysis and one whose analysis corresponded approximately to $C_4H_7ClO_2Pd$, as well as others under different conditions. They noted that these complexes decomposed in water at 100° to give HCl, metal, acrolein, and formaldehyde. The aldehydes were also detected as by-products in the formation of the complexes. Both complexes showed the presence of carbonyl bands at 1680 cm^{-1} in the infrared.

† Ziegler *et al.*[159,160] have used monosubstituted acetylenes, which gave complexes insoluble in water but soluble in organic solvents, for estimating Pd in the presence of large amounts of other metals.

Odaira *et al.*[164] reacted acetylene with $PdCl_2$ or $(PhCN)_2PdCl_2$ in acetic acid under pressure and obtained a black insoluble material which they suggested was a linear *trans*-polyacetylene. Small amounts of other organic products were also obtained.

A patent by Smidt *et al.*[99] reported that with $PdCl_2$ acetylene gave rise to butadiene, vinylacetylene, and even some divinylacetylene, the amounts depending on solvent and temperature.

These results, though somewhat fragmentary, suggest that (1) acetylene is readily oligomerized by $PdCl_2$; (2) that oxidation, either concurrent with oligomerization or subsequent, can take place; and (3) that a great deal of work under very carefully controlled conditions is still necessary before any conclusions whatsoever about either the complexes obtained or about possible mechanisms are drawn. The same is true in large measure for reactions of monosubstituted acetylenes. Again here the products are readily formed, as brownish complexes, which are usually of variable composition, and whose molecular weights change with time, possibly owing to autoxidation.[165] Maitlis *et al.*[165] were able to isolate highly colored organic tetramers (which may have been tetraphenyldihydropentalenes) and at least one trimer (2,4,6-triphenylfulvene) in very low yields from reactions of phenylacetylene with $PdCl_2$ under various conditions. However, the main products were uncharacterizable, nonstoichiometric, and noncrystalline complexes. On reduction, they gave moderately high molecular weight organic polymers. Similar results have been reported by Oshima[166] for phenylacetylene and by Babaeva and Kharitonov and their co-workers for some mono- and dihydroxy-acetylenes.[167–173] Attempts to duplicate some of the latter work, which claimed the formation of simple, crystalline monomeric complexes, were not successful.[165]

An interesting feature of these reactions is that benzenoid trimers were obtained only to a very small extent for a variety of acetylenes.[164, 165]

The most reliable and complete information concerns the disubstituted acetylenes. In 1959 Malatesta *et al.*[174] found that diphenylacetylene was converted into a crystalline complex (**I-43**) (R = Ph) of defined stoichiometry by palladium chloride in hydroxylic solvents; and that this complex was converted by acid to a tetraphenylcyclobutadienepalladium complex (**I-44**) (R = Ph, X = Cl). Further work by Maitlis *et al.*,[127,175–177] Vallarino and Santarella,[94] and others[131] and an X-ray structure determination by Dahl and Oberhansli[178] of the complex (**I-43**) (R = Ph, R′ = Et) obtained by Malatesta, has indicated that the following reaction occurred for diphenylacetylene and some *p,p′*-disubstituted diphenylacetylenes.†

† This reaction took place at ca. 20°; at higher temperatures, particularly in the presence of acid, deoxybenzoin ($PhCOCH_2Ph$) was the main product.[176]

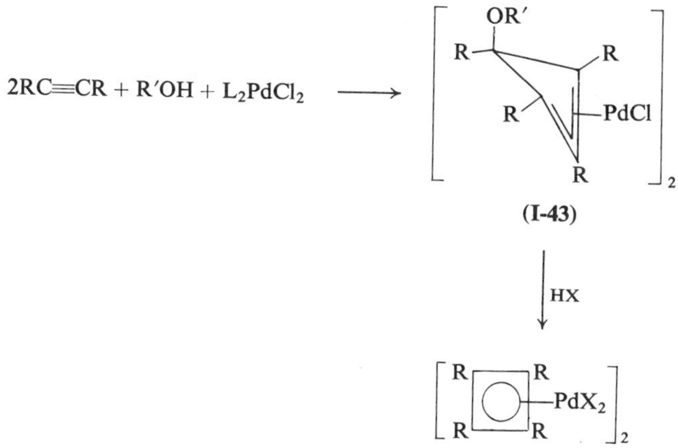

$$2RC{\equiv}CR + R'OH + L_2PdCl_2 \longrightarrow [\ldots PdCl]_2$$

(I-43)

↓ HX

$$[R_4C_4H \text{-} PdX_2]_2$$

(I-44)

(R = Ph, p-ClC_6H_4, p-MeC_6H_4)

The modification of this reaction, in which aprotic solvents (or solvent mixtures containing relatively small amounts of protonic solvents[179]) were used, led directly to hexaphenylbenzene and tetraphenylcyclobutadienepalladium chloride. The latter complex, $[Ph_4C_4(PdCl_2)_n]$, was unusual in that more than one $PdCl_2$ per R_4C_4 was present (see Volume I, Chapter IV, Section D,3). The exact value of n and also the relative amounts of dimer and trimer obtained depended on the conditions; more of the cyclobutadiene complex was obtained if the acetylene was added slowly to $(PhCN)_2PdCl_2$ in solution.[177] Under the opposite conditions, up to 20 equivalents of the acetylene per $PdCl_2$ could be trimerized.[175] Other p,p'-disubstituted diphenylacetylenes reacted similarly.[180] It was shown that tetraphenylcyclobutadienepalladium chloride was not an intermediate in the formation of hexaphenylbenzene.[175, 177]

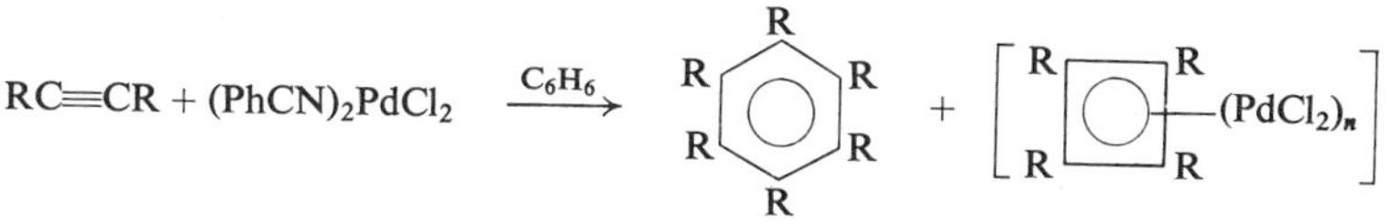

$$RC{\equiv}CR + (PhCN)_2PdCl_2 \xrightarrow{C_6H_6} C_6R_6 + [R_4C_4\text{-}(PdCl_2)_n]$$

(R = Ph, p-ClC_6H_4, p-MeC_6H_4, p-$MeOC_6H_4$)

None of these reactions appeared sensitive to air; oxidation and/or hydrolysis reactions did not occur if the reactions were conducted at ambient temperatures, and yields were good. There was no evidence for the occurrence of hydrogen-transfer reactions either. Palladium complexes with strongly or moderately strongly coordinating ligands (π-allylpalladium chloride, $bipyPdCl_2$, $[Bu_3PPdCl_2]_2$) were inactive in these reactions.[176]

Müller *et al.* have described an apparently similar reaction of *o*-di(phenylethynyl)benzene.[181]

Hosokawa and Moritani[182] and Avram *et al.*[183] showed that *t*-butyl(phenyl)acetylene reacted with $(PhCN)_2PdCl_2$ in benzene to give a 1,2-di-*t*-butyl-3,4-diphenylcyclobutadienepalladium chloride complex, which could be converted into the normal complex. No benzenoid trimers were reported.

$$t\text{-BuC}{\equiv}\text{CPh} + (\text{PhCN})_2\text{PdCl}_2 \xrightarrow{C_6H_6} [(t\text{-BuC}_2\text{Ph})_2(\text{PdCl}_2)_2] \xrightarrow{H_2O/DMSO,\ HCl} [(\text{Ph}_2t\text{-Bu}_2\text{C}_4)\text{PdCl}_2]_2$$

In contrast, di-*t*-butylacetylene did not oligomerize and only gave an acetylene complex, $[(t\text{-BuC}_2t\text{-Bu})\text{PdCl}_2]_2$.[122]

Two other unsymmetrical disubstituted acetylenes have been reacted with $PdCl_2$. Zingales[184] reported that under acidic conditions ethylphenylacetylene (1-phenyl-1-butyne) reacted with $PdCl_2$ in methanol to give 1,2,4-triethyl-3,5,6-triphenylbenzene and two complexes containing a tetramer of the acetylene as ligand, which were formulated as $(PhC_2Et)_4PdCl$ and $(PhC_2Et)_4Pd_2Cl_2$. Although the *un*symmetrical benzene was isolated, the tetramer complexes were suggested to be derived from the symmetrical 1,3,5,7-tetraethyl-2,4,6,8-tetraphenylcyclooctatetraene, and no isomers of these complexes were reported. Structures were proposed, but a full reexamination of the complexes using modern spectroscopic techniques would appear highly desirable.

Dietl and Maitlis[185] showed that in methylene chloride or chloroform, methylphenylacetylene (1-phenyl-1-propyne) was catalytically trimerized by $(PhCN)_2PdCl_2$. The major products were 1,2,4-trimethyl-3,5,6-triphenylbenzene (58 %) and 1,3,5-trimethyl-2,4,6-triphenylbenzene (39 %). A 3 % yield of 1,2,3-trimethyl-4,5,6-triphenylbenzene was also obtained from the mother liquors.

$$\text{PhC}{\equiv}\text{CMe} + (\text{PhCN})_2\text{PdCl}_2 \xrightarrow{CH_2Cl_2} \underset{(58\%)}{\text{1,2,4-Me}_3\text{-3,5,6-Ph}_3\text{C}_6} + \underset{(39\%)}{\text{1,3,5-Me}_3\text{-2,4,6-Ph}_3\text{C}_6} + \underset{(3\%)}{\text{1,2,3-Me}_3\text{-4,5,6-Ph}_3\text{C}_6}$$

$$\downarrow$$

$$[(\text{PhC}_2\text{Me})_3\text{PdCl}_2]_2 \xrightarrow[CDCl_3\ \text{or}\ PPh_3]{CH_2Cl_2\ \text{or}} \text{1,2,4-Me}_3\text{-3,5,6-Ph}_3\text{C}_6 + \text{1,3,5-Me}_3\text{-2,4,6-Ph}_3\text{C}_6$$

When the reaction was carried out in benzene, a 60% yield of a complex, $[(PhC_2Me)_3PdCl_2]_2$, was obtained in addition to the three benzenoid trimers. This complex was very labile and decomposed to the benzenoid trimers in $CDCl_3$ or CH_2Cl_2 solution, or on treatment with PPh_3. It, or a closely related complex, was hence a precursor to the benzenoid trimers, and therefore a mixture of several isomers.

Later Maitlis and his collaborators[186–188] found that an analogous complex, $[Cl(MeC_2Me)_3PdCl]_2$, was obtained in 50% yield from dimethylacetylene and $(PhCN)_2PdCl_2$ in benzene at 10°. Tetramers of the acetylene were also formed. The structure **(I-45)** was proposed for the complex $[Cl(MeC_2Me)_3PdCl]_2$ on spectroscopic and chemical evidence.

The route by which the complex **(I-45)** was formed was established by low-temperature NMR studies.[187] It involved primary formation of a π-acetylene complex, $Me_2C_2(PdCl_2)_2$(solv.), followed by a fast reaction, during which no other intermediates were detected, to give a complex, $[(Me_2C_2)_3(PdCl_2)_2]_2$, believed to be very closely related to **(I-45)**, and for which the structure **(I-46)** was proposed.

$$MeC{\equiv}CMe + (PhCN)_2PdCl_2 \underset{}{\overset{-50°}{\rightleftharpoons}} MeC_2Me(PdCl_2)_2solv.$$

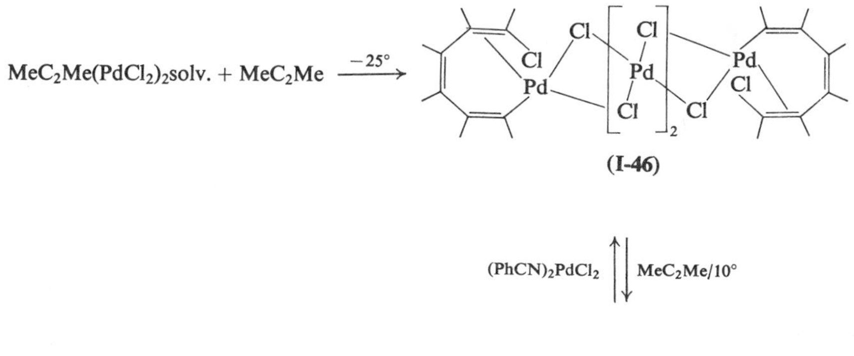

$$MeC_2Me(PdCl_2)_2solv. + MeC_2Me \xrightarrow{-25°}$$ **(I-46)**

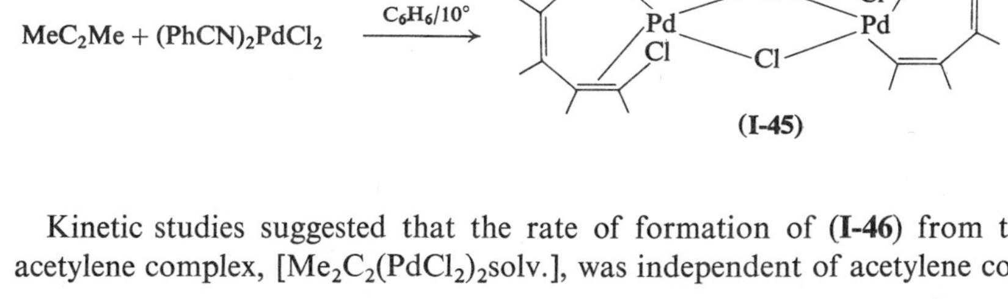

(PhCN)$_2$PdCl$_2$ ⇅ MeC$_2$Me/10°

$$MeC_2Me + (PhCN)_2PdCl_2 \xrightarrow{C_6H_6/10°}$$ **(I-45)**

Kinetic studies suggested that the rate of formation of **(I-46)** from the acetylene complex, $[Me_2C_2(PdCl_2)_2solv.]$, was independent of acetylene concentration. This is consistent with a rearrangement of the acetylene complex, such as a cis-ligand insertion into the Pd–Cl bond, as the rate-determining step.

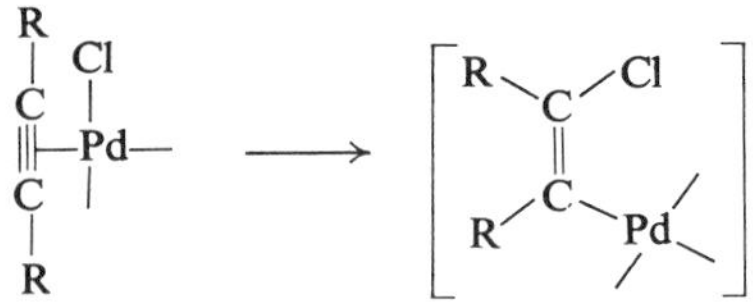

The σ-vinyl intermediate can then react rapidly with more acetylene in a fast cis insertion to give **(I-47)** (not isolated).

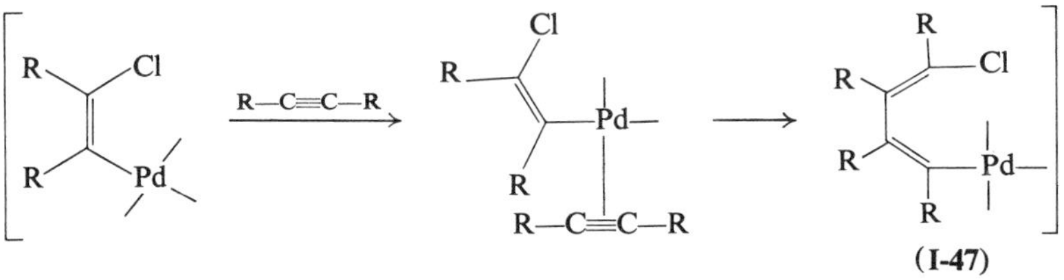

(I-47)

In the most general case, assuming that all disubstituted acetylenes behave similarly, **(I-47)** can react in two ways. When the acetylenic substituents are bulky (phenyl, *t*-butyl) further acetylenes complex and insert only with difficulty considerable or even exclusive cyclization, possibly via a cyclobutenyl complex, to a π-cyclobutadiene complex occurs. In fact, cyclobutadiene complexes are only obtained from acetylenes with two very bulky substituents.

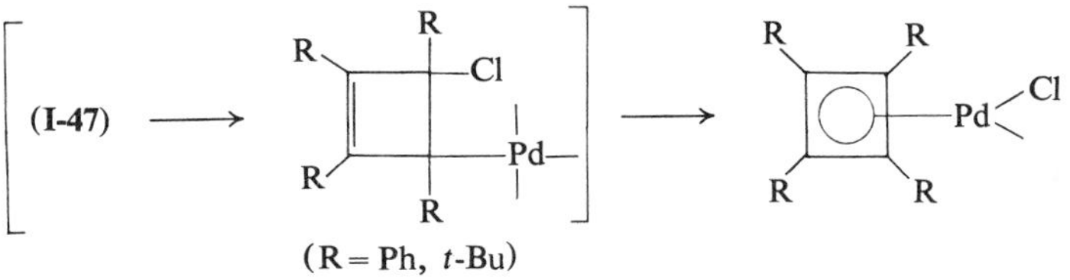

(R = Ph, *t*-Bu)

Alternatively, for less bulky substituents (methyl) a further reaction with an acetylene in the same sense can occur, leading to complexes analogous to **(I-45)** or **(I-46)**. Little is known about the mode by which the cyclization

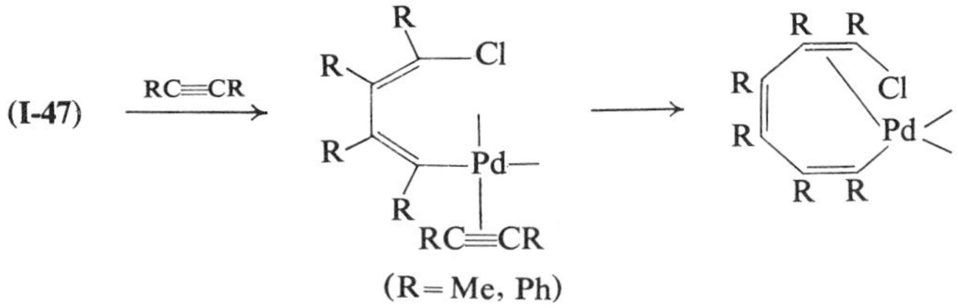

(R = Me, Ph)

reactions occur to give either the cyclobutadiene complex from **(I-47)** or the hexasubstituted benzene (R_6C_6) from **(I-45)** or **(I-46)**. In the latter case, however, the "allowed" electrocyclic disrotatory cyclization does not appear to take place (see below).[187,188]

In the dimethylacetylene reactions studied, and in the diphenylacetylene reactions too, there is evidence for the participation of two (or more) $PdCl_2$ units at each reactive site. The significance of this is not clear and it may simply be a phenomenon associated with the relative stability of $(PdCl_2)_n$ species compared to $PdCl_2$ in unpolar solvents in the absence of chloride, and not have any more profound mechanistic implications.

The mechanism can also be applied to explain the isolation of *endo*-alkoxytetraphenylcyclobutenylpalladium(II) complexes **(I-43)** from reactions of diphenylacetylene (and analogous acetylenes) with $PdCl_2$ in alcohols. In this case the reactive species could be an alkoxy-Pd(II) complex,

$$R'OH + PdCl_2 \rightarrow \text{"}R'OPdCl\text{"} + HCl$$

"R'OPdCl" + RC≡CR ⟶ [π-complex] ⟶ [R(OR')C=C(R)Pd] —RC₂R→ [alkyne-coordinated vinyl-Pd] ⟶ [butadienyl-Pd with OR'···Pd] ⟶ [cyclobutenyl-Pd] ⟶ [endo-alkoxycyclobutenyl-Pd]

(R = Ph, $p\text{-}ClC_6H_4$ or $p\text{-}MeC_6H_4$)

Alternatively, as this reaction occurs in a highly polar medium, OR^- could attack an acetylene π complex; this would lead to the *trans*-alkoxybutadienyl complex,

[RC≡CR π-complex with Pd] —OR'⁻→ [R'O(R)C=C(R)Pd] —RC₂R→ [alkoxybutadienyl-Pd]

(R = Ph, $p\text{-}MeC_6H_4$, $p\text{-}ClC_6H_4$)

which could then undergo a stereospecific conrotatory electrocyclic cyclization to give the observed *endo*-alkoxycyclobutenyl complex.

The trimer complexes $[Cl(Me_2C_2)_3PdCl]_2$, $[Cl(Me_2C_2)_3PdCl \cdot PdCl_2]_2$, and $[Cl(PhC_2Me)_3PdCl]_2$ all gave the hexasubstituted benzenes either on heating or in chlorinated solvents at 30°.

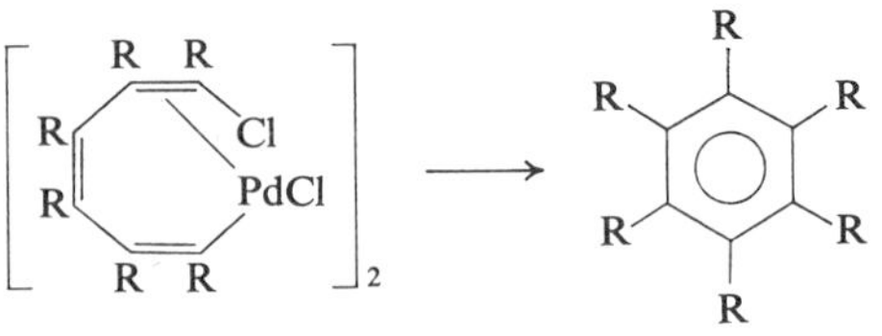

The complex from $CD_3C{\equiv}CCH_3$ gave hexamethylbenzene-d_9 which contained $10.2 \pm 0.5\%$ of 1,2,3-tris(trideuteromethyl)-4,5,6-trimethylbenzene.[188, 189]

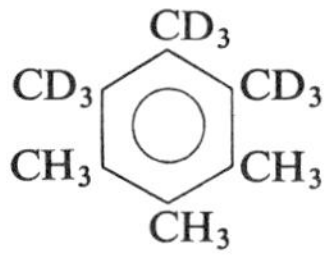

This result, together with the isolation of 1,2,3-trimethyl-4,5,6-triphenylbenzene from the trimerization of methyl(phenyl)acetylene[185] suggested that a scrambling reaction occurred during the cyclization of the acyclic trimer complex to the benzene. The NMR spectrum of **(I-45)** at 60° indicated that the complex was undergoing some kind of fluxional behavior and lent support to this suggestion.[187, 188]

The trimer complex $[Cl(Me_2C_2)_3PdCl]_2$, **(I-45)**, also reacted with triphenylphosphine, -arsine, and -stibine to give vinylpentamethylcyclopentadiene **(I-48)**, (α-chlorovinyl)pentamethylcyclopentadiene **(I-49)**, as well as some hexamethylbenzene. At low temperatures, vinylpentamethylcyclo-

$[Cl(MeC_2Me)_3PdCl]_2 + Ph_3E \longrightarrow$ + + $+ (Ph_3E)_2PdCl_2$

(I-45) **(I-48)** **(I-49)**

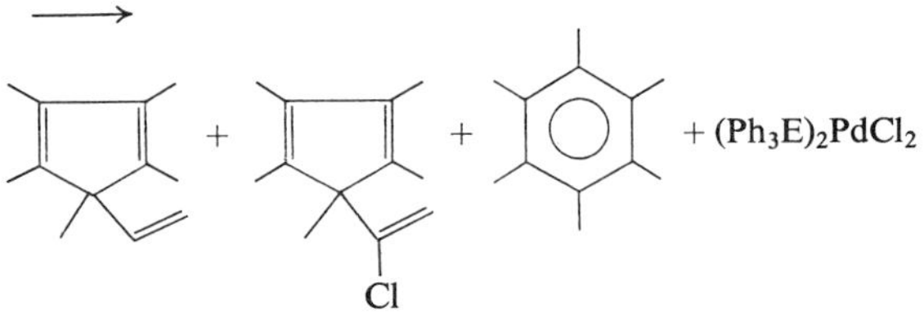

$(E = P, As, Sb)$

pentadiene was the major product and formation of **(I-49)** was completely suppressed. The NMR study of this reaction showed it to be very complex and to involve the formation of a fluxional intermediate.[188] The reaction in

Scheme I-3 was proposed to explain the observations. The key steps are as follows.

1. The complex **(I-45)** can exist in either of two conformations with the coordinated double bond perpendicular **(I-45a)** or coplanar **(I-45b)** to the coordination plane of the metal. The latter conformer is well suited to undergo "insertion" of the coordinated double bond into the Pd–C σ bond.

2. This reaction, assisted by L (PPh_3 or $AsPh_3$) occurs very readily (<–60°), the first observed product being **(I-50)**.

3. At higher temperatures (–25°) **(I-50)** undergoes a *reversible* valence tautomerism in which all the five methyls on the C_5 ring become equivalent, but where the stereochemistry about C-6 remains fixed.

4. At still higher temperatures (ca. 10°) decomposition of the intermediates to hexamethylbenzene and vinylpentamethylcyclopentadiene occurs.

5. Triphenylphosphine, but *not* triphenylarsine, is a sufficiently good base (in the Brønstead–Lowry sense) when present in excess to remove a proton from **(I-50)** giving the complex **(I-51)** and triphenylphosphonium chloride. This complex also decomposes above ca. –10° to give hexamethylbenzene and **(I-48)**, probably via regeneration of **(I-50)**.

Many other reagents reacted with $[Cl(Me_2C_2)_3PdCl]_2$, **(I-45)**, to give derivatives of pentamethylcyclopentadiene[188]:

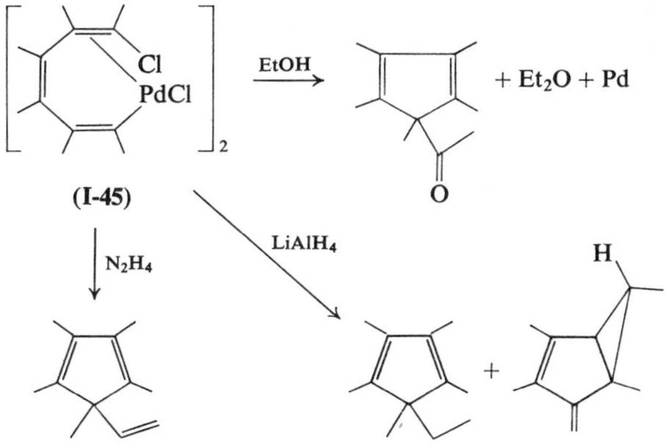

Cl PdCl]$_2$ —EtOH→ + Et_2O + Pd

(I-45)

N_2H_4 LiAlH$_4$ H

These reactions presumably proceed by analogous paths. In contrast, halogens, acids, and a few other reagents gave hexamethylbenzene with **(I-45)**.[188]

Although monosubstituted acetylenes generally reacted with palladium chloride to give mixtures which were difficult to resolve. Avram *et al.* found that *t*-butylacetylene gave simpler products. In acetone the acetylene was trimerized to 1,3,5-tri-*t*-butylbenzene, while in benzene, the complex $[(t\text{-}BuC_2H)_3PdCl_2]_n$ was obtained. The complex was assumed to be monomeric and these workers

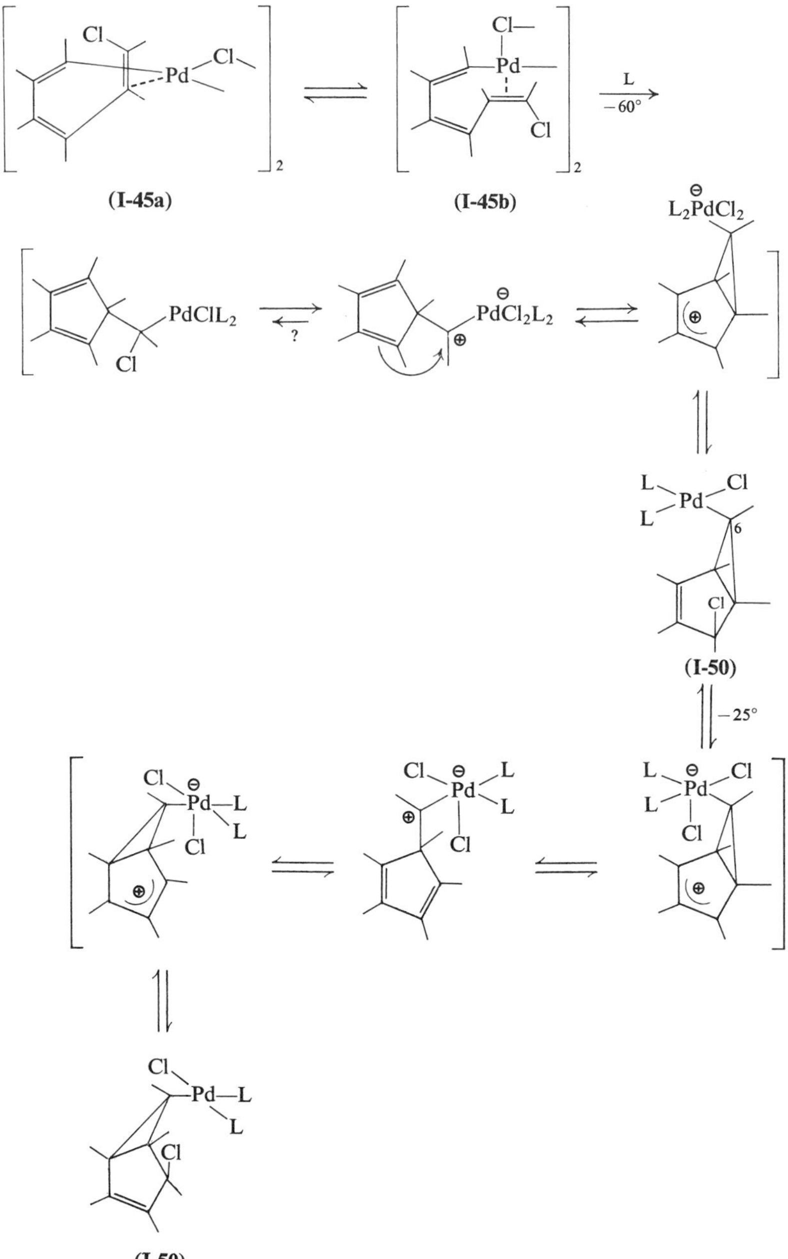

(I-45a) (I-45b)

(I-50)

(I-50)

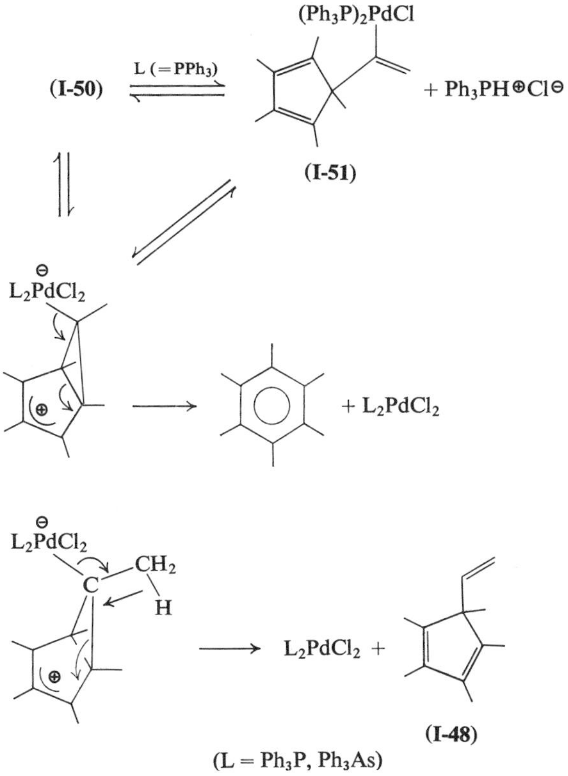

$(Ph_3P)_2PdCl$

(I-50) $\underset{}{\overset{L\,(=PPh_3)}{\rightleftharpoons}}$ + $Ph_3PH^{\oplus}Cl^{\ominus}$

(I-51)

$L_2PdCl_2^{\ominus}$ ⟶ + L_2PdCl_2

$L_2PdCl_2^{\ominus}$ CH2 C H ⟶ L_2PdCl_2 +

(I-48)

($L = Ph_3P, Ph_3As$)

SCHEME I-3

proposed it to be a Dewar benzene complex, 1,2,5-tri-*t*-butylbicyclo[2.2.0]-hexadienepalladium chloride.[190] A reinvestigation by Kaiser and Maitlis showed that the complex was dimeric, i.e., $[(t\text{-}BuC_2H)_3PdCl_2]_2$, and was *not* a Dewar benzene complex. The two halogens were also nonequivalent.[191] The structure is not known, but it may be closely related to **(I-45)**; however, all attempts to obtain a benzenoid trimer from the complex have failed. Avram *et al.* have described a number of degradations to organic materials but, on the basis of the published evidence, the structures suggested are not proven. These workers also reported complexes $[(\text{acetylene})_3PdCl_2]_n$ (*n* not determined) from dimethyl acetylenedicarboxylate and other acetylenes, which were assigned Dewar benzene structures without supporting evidence.[192]

Some Ni(II) and Pt(II) compounds reacted under some conditions with acetylenes to give cyclobutadiene complexes and trimers[193, 194] (see Volume I, Chapter IV, Section D,3).

Bryce-Smith has found that certain types of palladium on charcoal catalysts were active for the trimerization of dimethyl acetylenedicarboxylate and the polymerization of some monosubstituted acetylenes.[195]

Moseley and Maitlis[196] observed an apparently related reaction of bis-(dibenzylideneacetone)palladium(O) [$(DBA)_2Pd$][197] with dimethyl acetylene-dicarboxylate. The first isolated complex was the palladiacyclopentadiene (**I-51a**), which was probably a tetramer. On heating with more acetylene hexamethyl mellitate was formed; other reactions are summarized below,

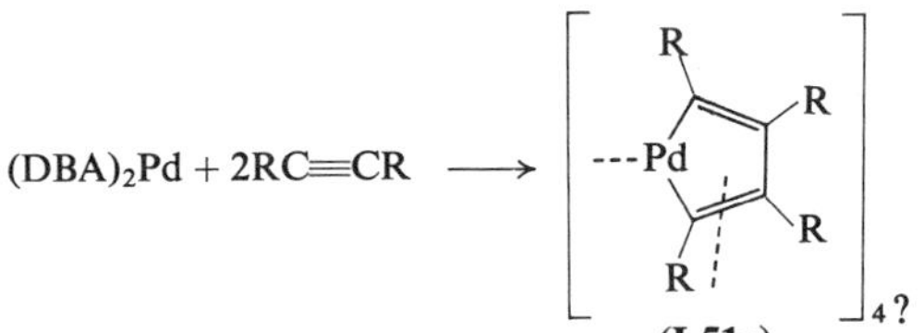

$$(DBA)_2Pd + 2RC{\equiv}CR \longrightarrow [\,\text{(I-51a)}\,]_{4?}$$

(I-51a)

(I-51a) $\xrightarrow{Br_2}$ RC(Br)=C(R)–C(R)=C(Br)R

(I-51a) $\xrightarrow{2L}$ $L_2Pd(C_4R_4)$

(I-51a) $\xrightarrow{PhC_2Ph}$ $C_6R_4Ph_2$

(I-51a) $\xrightarrow{RC_2R/60^\circ}$ C_6R_6

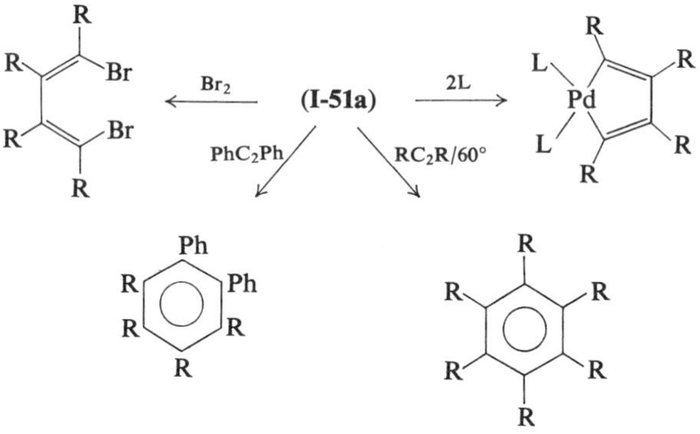

(R = COOMe, L = Ph_3P, Cl^-, PhCN, ½(bipy), etc.)

e. *Miscellaneous*

Sakai *et al.*[198, 199] noted that a $PdCl_2$–$CuCl_2$ catalyst was very active for olefin–formaldehyde condensation using branched-chain olefins, e.g.,

$$\mathrm{Me_2CHCH{=}CH_2 + HCHO \xrightarrow{PdCl_2\text{–}CuCl_2}}$$ (4,4,5-trimethyl-1,3-dioxane) + (4-isopropyl-1,3-dioxane)

The yields were much higher than for the conventional H^+-catalyzed (Prins) reaction.

Atkins *et al.*[200] and Hata *et al.*[201] found that a variety of functionally substituted allyls reacted with β diketones and similar compounds containing an activated methylene group in a Pd catalyzed reaction; for example,

$$\mathrm{CH_2{=}CHCH_2X + MeCOCH_2COMe \xrightarrow{Pd(acac)_2/PPh_3}}$$

$$\mathrm{CH_2{=}CHCH_2CH(COMe)_2 + (CH_2{=}CHCH_2)_2C(COMe)_2 + HX}$$

$$(\mathrm{X = OH, OPh, OAc, NEt_2})$$

Isonitriles will also insert into Pd–C bonds[202]; both reactions with π-allylic and with Pd-methyl complexes are known. In the latter case a stepwise oligomerization can also occur.

$$\mathrm{[(\pi\text{-}C_3H_5)PdCl]_2 + RNC \longrightarrow [CH_2{=}CHCH_2C({=}NR)(RNC)PdCl]_2 \xrightarrow{EtOH/Na_2CO_3} CH_2{=}CHCH_2C({=}NR)OEt}$$

$$\mathrm{L(Me)PdI(L) + RNC \longrightarrow [Me\text{–}C({=}NR)(L)PdI]_2 \xrightarrow{RNC/L} \xrightarrow{RNC}}$$

(cyclic product: RN=C–C(=NR)–Pd(L)(I)–N(R)=C(Me)–, five-membered ring)

$$(\mathrm{L = MePh_2P;\ R = C_6H_{11}})$$

D. OXIDATIVE COUPLING

A very important series of reactions are those in which two hydrocarbons are coupled, usually with loss of two hydrogens, in a stoichiometric reaction with $PdCl_2$.

$$2RH + PdCl_2 \rightarrow R\text{–}R + 2HCl + Pd$$

The importance of these reactions arises from the fact that the palladium metal can again be reoxidized, say by Cu(II), and thus the reaction becomes effectively catalytic in palladium. The Cu(I) produced can be reoxidized by oxygen and in this way the reaction can be made catalytic in both metals.

A large number of variations on this basic theme are known, some of which are quite exotic and for which little mechanistic evidence is currently available.

The key to understanding these reactions can be found in the observation by Heck[19, 203] and Henry[45] that arylmercury compounds react with palladium chloride, in the absence of other reactive substrates such as olefins or CO, to give biaryls.

$$2ArHgX + PdCl_2 \rightarrow Ar\text{–}Ar + 2HgXCl + Pd$$

Presumably the organomercury compound acts as an arylating agent toward the palladium giving a monoarylpalladium species, but the actual mechanism of the coupling reaction is not known. Since free radicals are not usually involved in such reactions, diarylpalladium or possibly monoarylpalladium cluster complexes may be intermediates.

In this connection, Maitlis and Stone[204] showed that dimethyl(bipyridyl)-palladium gave ethane on treatment with C_3F_7I,

$$bipyPdMe_2 + C_3F_7I \rightarrow bipyPd(C_3F_7)_2 + C_2H_6$$

and Yamamoto and Ikeda[205] found that diethyl(bipyridyl)nickel gave *n*-butane on reaction with acrylonitrile at 25°.

$$bipyNiEt_2 + CH_2{=}CHCN \rightarrow n\text{-}C_4H_{10} + C_2H_6 + C_2H_4 + bipyNi(CH_2{=}CHCN)_2$$

Rausch and Tibbetts[206] have noted that reaction of $bipyPdCl_2$ with phenyllithium gave only biphenyl and no complex was isolated.

Organocopper complexes, $(RCu)_n$, readily give R–R[207–211]; Cairncross and Sheppard[212] have studied the decomposition of $[m\text{-}CF_3C_6H_4Cu]_8$ to $(m\text{-}CF_3C_6H_4)_2$ and copper and have found this to occur stepwise in a unimolecular process.

$$R_8Cu_8 \rightarrow R_2 + R_6Cu_8 \qquad (R = m\text{-}CF_3C_6H_4\text{–})$$
$$R_6Cu_8 \rightarrow R_2 + R_4Cu_8$$
$$R_4Cu_8 \rightarrow R_2 + R_2Cu_8$$
$$R_2Cu_8 \rightarrow R_2 + Cu_8$$

The palladium(II)-promoted coupling reaction was first reported by Hüttel *et al.*,[213, 214] who found that α-substituted styrenes were coupled by $PdCl_2$ in acetic acid.

$$2\ (Ph)(R)C{=}CH_2 \longrightarrow (Ph)(R)C{=}CH{-}CH{=}C(Ph)(R)$$

(R = Me, Ph)

This reaction also proceeded in the presence of base. Oxidation products such as PhCOR were also obtained. Volger[215, 216] has investigated this reaction in some detail using palladium acetate and sodium acetate in acetic acid. Deuteration studies, using both $PhC(CD_3){=}CH_2$ and DOAc, showed that the former incurred no loss of label on coupling, and that no deuterium was incorporated when the reaction on $PhC(CH_3){=}CH_2$ was carried out in DOAc.

In the absence of acetate ion, a *nonoxidative* coupling occurred to give 4-methyl-2,4-diphenyl-2-pentene.

$$(Ph)(Me)C{=}CH_2 \xrightarrow{Pd(OAc)_2} MeC(Ph){=}CHC(Me)_2Ph$$

Volger suggested that a bimolecular π-olefin complex, such as **(I-52)**, is involved in the oxidative coupling. Here the two CH_2 groups are in close proximity and the abstraction of H from $CH_2{=}$ by the metal is facilitated by the presence of acetate ion. In a synchronous process the H^- abstracted is oxidized

(I-52): two $CH_2{=}C(Ph)Me$ units, each π-bonded to Pd(OAc)₂, with bridging OAc groups between the two Pd atoms $\longrightarrow$

$$(Ph)(Me)C{=}CHCH{=}C(Ph)(Me) + 2HOAc + Pd(OAc)_2 + Pd^0$$

to H^+ by the metal and two electrons are transferred to the vinylic group via the metal creating a vinylic carbanion, which, in turn, then attacks the other adjacent coordinated olefin. H^- is transferred from this onto the metal, to give H^+, Pd^0, and 2,5-diphenyl-2,4-hexadiene.

It is not necessary, however, for a vinylic carbanion to be created; oxidative addition of PhMeC:CH–H to palladium acetate followed by a reductive elimination could give a σ-vinylic palladium(II) acetate and acetic acid. Alternatively, this reaction can be regarded as an electrophilic substitution (palladation) of a vinylic hydrogen. The σ-vinylic complex can then coordinate a further molecule of the styrene and give the 2,4-hexadiene after a cis insertion followed by elimination of HPdOAc.

$$PhMeC{=}CH_2 + Pd(OAc)_2 \rightarrow PhMeC{=}CHPd(OAc) + HOAc$$

$$PhMeC{=}CH_2 + PhMeC{=}CHPd(OAc) \rightarrow PhMeC{=}CH{-}Pd(OAc)(CH_2{=}CPhMe) \rightarrow$$

$$PhMeC{=}CH \cdot CH_2C(PdOAc)PhMe \rightarrow PhMeC{=}CHCH{=}CMePh + Pd^0 + HOAc$$

In the presence of $PdCl_2$, 2,5-diphenyl-2,4-hexadiene was slowly isomerized to 2,5-diphenyl-1,5-hexadiene. A second, much slower, oxidative coupling also occurred with the 2,4-hexadiene to give *p*-terphenyl.[216]

Ph–C(Me)=CH–CH=C(Me)–Ph —$PdCl_2$→ Ph–C(=CH_2)–CH_2CH_2–C(=CH_2)–Ph

Ph–C(Me)=CH–CH=C(Me)–Ph —$Pd(OAc)_2$ (slow)→ Ph–C₆H₄–Ph

Kohll and van Helden[217,218] found that vinyl acetate reacted with palladium acetate to give a variety of products including the coupled product, *trans,trans*-1,4-diacetoxy-1,3-butadiene.

$$CH_2{=}CHOAc \rightarrow AcOCH{=}CHCH{=}CHOAc + (AcO)_2CHCH{=}CHCH_2OAc + (AcO)_2CH(CH_2)_2CH(OAc)_2 + AcH + Ac_2O + CH_2CHCH{=}CHOAc$$

This reaction could be made catalytic in palladium using copper acetate. A mechanism similar to the one proposed by Volger was suggested to account for the formation of the diacetoxybutadiene.

The other dimer products could also arise from a vinyl acetate complex, for example,

$$AcOCH{=}CH_2 + Pd(OAc)_2 \longrightarrow \underset{AcO}{\diagup}\!\!\Vert{-}Pd(OAc)_2 \xrightarrow{OAc^-}$$

$$(AcO)_2CHCH_2{-}Pd(OAc)_2^- \xrightarrow{CH_2{=}CHOAc} (AcO)_2CHCH{-}Pd(OAc)_2^- \cdots \Vert{-}OAc \longrightarrow$$

$$(AcO)_2CHCH_2CH_2\underset{\displaystyle OAc}{\underset{|}{C}}H{-}Pd(OAc)_2^- \xrightarrow{H\ shift} (AcO)_2CHCH_2\underset{\displaystyle CH_2{-}OAc}{\underset{|}{C}}H{-}Pd(OAc)_2^- \longrightarrow$$

$$(AcO)_2CHCH{=}CHCH_2OAc + Pd^0 + OAc^- + HOAc$$

These mechanisms involve either direct nucleophilic attack by acetate on the complexed olefin, or a cis-ligand insertion accompanied by attack of acetate on the metal, and contrast with Volger's mechanism in which acetate "activates" a vinylic hydrogen to displacement by the metal.

In the presence of chloride ion, however, vinyl acetate gave only acetaldehyde and acetic anhydride. Little information on the generality of this coupling reaction has so far been published.

A reaction where more information is available is the oxidative coupling of arenes to biaryls, catalyzed by Pd(II). This reaction was first disclosed by van Helden and Verberg of the Shell group in Amsterdam.[219] They found that the reaction

$$2RC_6H_5 + PdCl_2 + 2NaOAc \rightarrow RC_6H_4{-}C_6H_4R + Pd + 2NaCl + 2HOAc$$

did not proceed in the absence of acetate, or if palladium bromide or iodide were used in place of the chloride. The rate was found to be first-order with respect to $PdCl_2$ and RC_6H_5 but independent of $[OAc^-]$. This suggested that the rate-determining step was the formation of a complex between $PdCl_2$ and the arene. The rate constants did not change greatly with different R groups [$R = Me$ ($k_2 = 0.11$) > H (0.09) > Cl (0.07) > MeOOC (0.04)], but electron-releasing substituents did promote the reactions. The isomer distribution of biphenyls was in agreement with a mechanism involving aromatic electrophilic substitution, for example, toluene gave 4,4′-, 3,3′-, 3,4′-, and 3,2′-bitolyls in the ratio of 20:20:35:25.[219,220] van Helden and Verberg also suggested that phenyl free radicals were not involved since no phenol was formed in the presence of oxygen. Unger and Fouty[221] have reported on the temperature variation of the ratios of isomeric bitolyls obtained from coupling toluene under these conditions. At 25° the 2,4′-isomer predominated, whereas above 50° the 3,4′-isomer was the major product. Overall yields increased on heating, from 16% in 7 days at 25° to 54% in 3 hours at 110°. Other variables only had minor effects. On addition of mercuric acetate, complete conversion to bitolyls occurred and more of the 4,4′-isomer was formed, rising to a maximum of 70% for a Hg:Pd ratio of 2. This reaction was strongly catalyzed by perchloric acid. The reactions of *p*-tolylmercuric acetate carried out by these workers are discussed below.

Bryant *et al.*[222] found that the coupling of toluene to bitolyls was very dependent on the ratio of acetate:palladium. At high acetate ratios (20:1) in the *absence of chloride* the main products were not bitolyls but $PhCH_2OAc$ and $PhCH(OAc)_2$, arising from benzylic oxidation. They suggested that for coupling to occur an aggregate containing two or more Pd atoms joined by bridging ligands was necessary. At high acetate concentrations, these bridges were broken, and benzylic oxidation occurred. Chloride bridges were less easily broken than acetate bridges and hence, in the presence of chloride, coupling was the main reaction.

Davidson and co-workers[223–225] have investigated the effects of oxygen and other factors on the coupling reactions catalyzed by palladium acetate in chloride-free media. The effect of oxygen (60–100 atm) was remarkable in that acetoxylation both of benzene (to give phenylacetate) and toluene was repressed, and coupling, to biphenyl and 3,3′-bitolyl, respectively, was the only reaction observed. In the absence of oxygen, phenyl acetate (or benzyl acetate from toluene) was found. The inhibition of these latter reactions by oxygen was consistent with the scavenging of a free-radical intermediate in low concentration.

These coupling reactions were strongly catalyzed by perchloric acid (contrast the reaction of van Helden and Verberg, in the presence of chloride, described above). This suggested that an electrophilic substitution of metal for aromatic hydrogen was occurring and that the perchloric acid acted by increasing the electrophilicity of the palladium,

$$Pd(OAc)_2 + H^+ \rightarrow PdOAc^+ + HOAc$$

The reaction with benzene was found to be fast even below 60°. Metal was not precipitated, but the solution acquired a magenta color, shown to be due to a Pd(I) species, and a 50% yield of biphenyl was obtained. The kinetics of the reaction indicated it to be first-order in palladium acetate and in benzene; furthermore, the large primary isotope effect ($k_H/k_D \approx 5.0$) found by comparing benzene and hexadeuterobenzene was similar to that for mercuration reactions.

Davidson *et al.*, therefore, interpreted the reactions as involving electrophilic substitution of the benzene to give a phenylpalladium(II) species, which then decomposed in a concerted process, without the intermediacy of free phenyl radicals, to biphenyl and Pd(I). The Pd(I) complex did not give more biphenyl with benzene under these conditions. It was stable, in the absence of chloride, to disproportionation (see also below and Volume I, Chapter VI, Section C).

$$C_6H_6 + Pd(II) \longrightarrow PhPd(II) + H^+$$

$$2PhPd(II) \rightarrow Ph\text{—}Ph + 2Pd(I)$$

Aromatic electrophilic substitution reactions are quite common for a number of metals including mercury(II), thallium(III), and gold(III).

Additional evidence for this type of reaction comes from the observation by Cope and Siekman[226] that azobenzene reacts with Pd(II) and Pt(II) to give complexes in which aromatic substitution has effectively occurred (see Volume I, Chapter II, Section C,3,b,v).

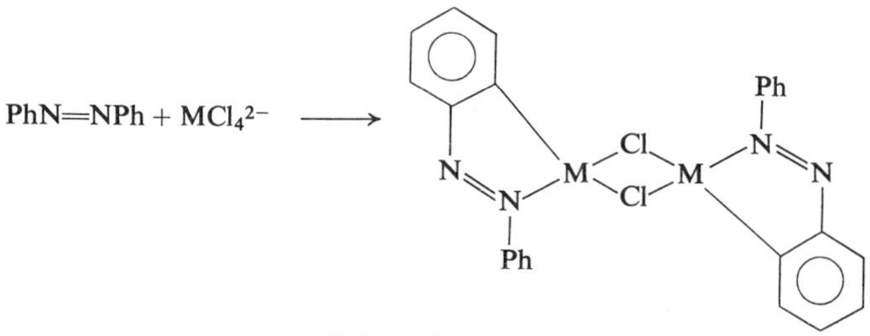

$$PhN{=}NPh + MCl_4^{2-} \longrightarrow$$

[M = Pd(II), Pt(II)]

Electrophilic substitution is also assisted by the presence on the metal of weakly bonded ligands such as acetate.

The reactions of palladium acetate with arylboron or arylmercury compounds have been studied by Davidson and Triggs[224] and Unger and Fouty.[221] The former workers found that *p*-tolylboronic acid [*p*-$MeC_6H_4B(OH)_2$] reacted with the acetate in acetic–perchloric acid to give only 4,4′-bitolyl and a Pd(I) complex. The reaction with *p*-tolylmercuric acetate was also acid-catalyzed, but bis(*p*-tolyl)mercury reacted instantly at 20° even without perchloric acid. However, only one *p*-tolyl group was consumed under these conditions, and the reaction was therefore represented as:

$$Ar_2Hg + Pd(OAc)_2 \rightarrow ArPdOAc + ArHgOAc$$

$$ArPdOAc + Ar_2Hg \rightarrow Ar_2 + ArHgOAc + Pd$$

(Ar = *p*-tolyl)

This result probably rules out the possibility that Ar_2Pd was the reactive intermediate. This reaction and the slow reaction of *p*-tolylmercuric acetate with palladium acetate in the absence of acid gave only 4,4′-bitolyl. In the presence of acid this isomer still predominated (70 %) in the latter reaction, but appreciable amounts of 2,4′- and 3,4′-bitolyls were also obtained. This was ascribed to the very high reactivity of ArPdOAc in acid which allowed it to attack other species, apart from ArHgOAc, such as toluene. However, it is presumably not impossible to exclude an acid-catalyzed isomerization of the *p*-tolylpalladium acetate intermediate to give some *o*- and *m*-tolylpalladium acetates.

It therefore appears that the acid-catalyzed coupling of aromatic hydrocarbons can be well explained on the basis of an electrophilic palladation, followed by a very fast concerted coupling of two aromatic groups and reduction of the metal to Pd(I). The base-catalyzed reactions of van Helden and Verberg also appear to involve a similar electrophilic palladation as primary step. It is possible that acetate has to be present here to convert the $PdCl_2$ to an acetato species (at least to a small extent) which is then the reactive reagent. The effect of excess acetate, found by Bryant *et al.*, is then to lower the electrophilicity of the palladium. It is instructive to compare the yields of isomeric bitolyls obtained by Davidson and Triggs[224] (acid-catalyzed, chloride free) and by Unger and Fouty[221] (acetate-catalyzed) in acetic acid at 50° (Table I-3).

TABLE I-3

Ratios of Isomeric Bitolyls Obtained from Toluene[a]

Reaction	2,2′-	2,3′-	2,4′-	3,3′-	3,4′-	4,4′-
Acid-catalyzed (Cl^- absent)	0.7	4.6	11.4	6.9	34.2	42.2
Base-catalyzed (Cl^- present)	Trace	15.3	27.2	11.6	28.4	17.4

[a] From Davidson and Triggs[224] and Unger and Fouty.[221]

Total yields are about the same (35, 31 %) based on Pd. The main difference is in the higher amount of 4-substitution which occurs in the acid-catalyzed reactions.

An analogy to this is the mercuration of toluene, where Brown and McGarry[227] found the ratio of *o*-:*m*-:*p*-tolylmercuric acetates to be 20.5:9.5:69.5 for the perchloric acid-catalyzed reaction and 30:13:56 for the uncatalyzed reaction in acetic acid at 50°. Both reactions proceeded by electrophilic mercuration, but in the former case the reagent was much more specific. Clearly, the same phenomenon occurs in the palladation reactions.

Sakakibara *et al.*[228] have noted that the base-catalyzed coupling reactions were repressed in favor of a new reaction, carboxylation, when conducted in the presence of acetic anhydride.

$$C_6H_6 + PdCl_2 + NaOAc + Ac_2O + AcOH \rightarrow \underset{(19\%)}{Ph_2} + \underset{(27\%)}{PhCOOH}$$

Direct carboxylation was also possible using sodium dimalonatopalladate[229];

$$CH_2(COOH)_2 + PdCl_2 + NaOAc \xrightarrow{HOAc/CCl_4} Na_2\left[\begin{array}{c} \text{O} \quad\quad\quad\quad \text{O} \\ \| \quad\quad\quad\quad \| \\ \text{C—O} \;\; \text{O—C} \\ CH_2 \quad \text{Pd} \quad CH_2 \\ \text{C—O} \;\; \text{O—C} \\ \| \quad\quad\quad\quad \| \\ \text{O} \quad\quad\quad\quad \text{O} \end{array}\right]$$

$$Na_2[\{CH_2(COO)_2\}_2Pd] + C_6H_6 \xrightarrow{HOAc/Ac_2O} PhCOOH \quad (98\%).$$

Nishimura *et al.*[230] have reported yet another variant of this reaction; when it was carried out using propionic, rather than acetic, acid and anhydride in the presence of sodium propionate, the main product from benzene was cinnamic acid together with a small amount of *trans*-stilbene.

$$C_6H_6 + PdCl_2 + NaOCOEt + (EtCO)_2O + EtCOOH \xrightarrow{100^\circ}$$

$$\underset{(28\%)}{PhCH{=}CHCOOH} + \underset{(3\%)}{PhCH{=}CHPh} + CO_2 + C_2H_4 + Pd$$

Palladium propionate reacted similarly. In the absence of propionic anhydride, cinnamic acid was a minor product, the major ones being biphenyl and β,β-diphenylacrylic acid. The reaction of benzene with sodium butyrate, butyric acid, and palladium chloride gave β-methylcinnamic acid and a little benzoic acid. In the presence of butyric anhydride, benzoic acid, propylene, and carbon dioxide were formed.

$$C_6H_6 + PdCl_2 + NaOOCC_3H_7 + C_3H_7COOH \rightarrow PhCMe{=}CHCOOH + PhCOOH$$

$$\xrightarrow[(C_3H_7CO)_2O]{} PhCOOH + CH_2{=}CHMe + CO_2$$

In both the carboxylation of arenes and the formation of the cinnamic acids, electron-releasing substituents (Me, MeO) directed substitution to the para position; however, highest yields were obtained from benzene itself. Electron-withdrawing substituents (NO_2, Ac) on the benzene in the carboxylation reaction directed meta and gave low yields. For the moment, in the absence of further details, no mechanisms can be suggested, though it is possible that the reactions with propionate and butyrate may go via primary formation of the acrylic acid which is then phenylated. In fact, Sakakibara *et al.*[231] have very recently shown that benzene reacted with acrylic acid, in the presence of $PdCl_2$ and sodium acetate, to give cinnamic acid.

A further reaction involving oxidative coupling is the phenylation of olefins described by Moritani and Fujiwara and their collaborators.[232–240] Various forms of the reaction are known, including:

$$[\text{PhCH=CH}_2\text{—PdCl}_2]_2 + RC_6H_5 \xrightarrow{\text{AcOH}} R\text{—}C_6H_4\text{—CH=CH—}C_6H_5$$

$$p\text{-}RC_6H_4CH{=}CH_2 + R'C_6H_4 \xrightarrow[\text{Pd(OAc)}_2\text{/AcOH}]{\text{PdCl}_2\text{/OAc}^- \text{ or}} R\text{—}C_6H_4\text{—CH=CH—}C_6H_4\text{—}R'$$

The last reaction is the most efficient and can be made catalytic in palladium in the presence of copper acetate† and oxygen (55 atm) to give *trans*-stilbene from benzene and styrene, together with a small amount of 1,4-diphenylbutadiene by oxidative coupling of styrene.

Steric effects are very important as shown by the decrease in yield as the olefin becomes more heavily substituted.

$$PhCH{=}CH_2 \rightarrow trans\text{-}PhCH{=}CHPh \quad (95\%)$$

$$Ph_2C{=}CH_2 \rightarrow Ph_2C{=}CHPh \quad (72\%)$$

$$trans\text{-}PhCH{=}CHPh \rightarrow Ph_2C{=}CHPh \quad (28\%)$$

$$Ph_2C{=}CHPh \rightarrow Ph_2C{=}CPh_2 \quad (13\%)$$

In competitive reactions with styrene, benzene was the most reactive and substituted benzenes showed the following partial rate factors[239]:

Substituent	ortho	meta	para
H (benzene)			1.0
Me	0.11	~0	4.17
Et	~0	~0	3.91
Cl	0.13	0.44	2.50
NO_2	~0	1.50	~0

These results indicate that the substituent on benzene controls the orientation of substitution but *not* the rate to any great extent,‡ and are consistent with a two-stage reaction path, involving electrophilic palladation of the aromatic ring followed by oxidative coupling with the olefin. Danno *et al.*[242] found no evidence for a hydride shift in the phenylation of styrene.

† However, not all oxidizing cocatalysts act in the same way; for example, reaction of benzene and $[C_2H_4PdCl_2]_2$ in the presence of silver nitrate only gave biphenyl.[241] Similar results were observed for complexes of other olefins and for substituted benzenes.

‡ Naphthalene reacted with styrene to give 2-*trans*-styrylnaphthalene and none of the 1-isomer. This may be due to steric factors.

$$RC_6H_5 + Pd(OAc)_2 \rightarrow [RC_6H_4PdOAc] + HOAc$$

$$\left[RC_6H_4PdOAc \xrightarrow{PhCH{=}CH_2} RC_6H_4{-}PdOAc(CH_2{=}CHPh) \longrightarrow RC_6H_4CH_2CH(Ph){-}PdOAc\right] \longrightarrow$$

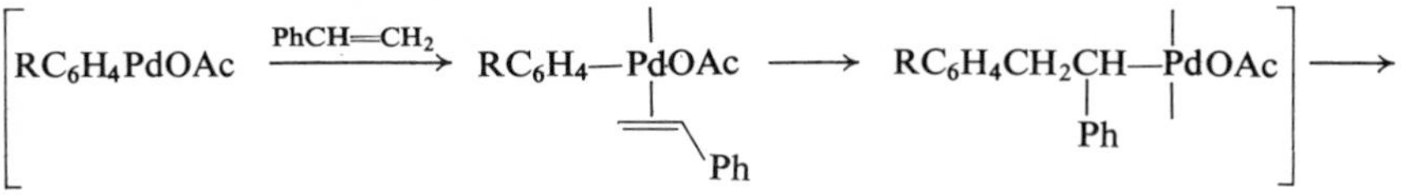

$$RC_6H_4CH{=}CHPh + HOAc + Pd^0$$

This reaction is now apparently analogous to the olefin arylation using arylmercury(II) complexes and Pd(II) reported by Heck and discussed in Section A. In this case, however, the palladium(II) substitutes the arene directly and an external arylating agent is not necessary.

Since arylpalladium(II) complexes are very unstable and react rapidly with olefins, they are unlikely to be present in large amount. The rate-determining step, therefore, presumably occurs during the palladation reaction. To explain the partial rate factors observed for the substituted benzenes it is necessary to assume that the slow step here is the formation of a π-arene complex **(I-53)** which then decomposes rapidly to the σ-complex intermediate **(I-54)**. The

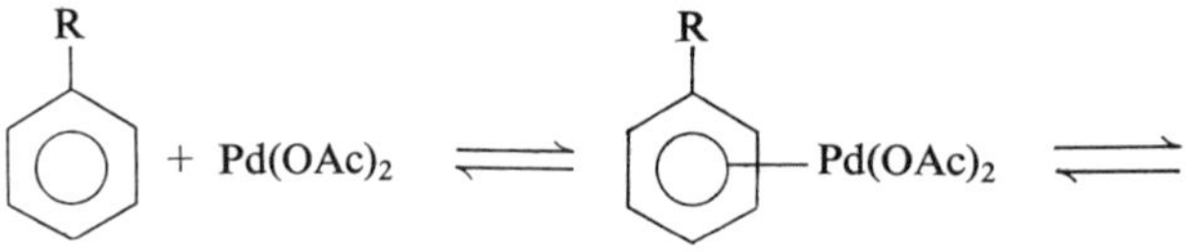

$$RC_6H_5 + Pd(OAc)_2 \rightleftharpoons RC_6H_5{-}Pd(OAc)_2 \rightleftharpoons$$

(I-53)

$$[RC_6H_5^{\oplus}(H)Pd(OAc)_2^{\ominus}] \longrightarrow RC_6H_4Pd(OAc) + HOAc$$

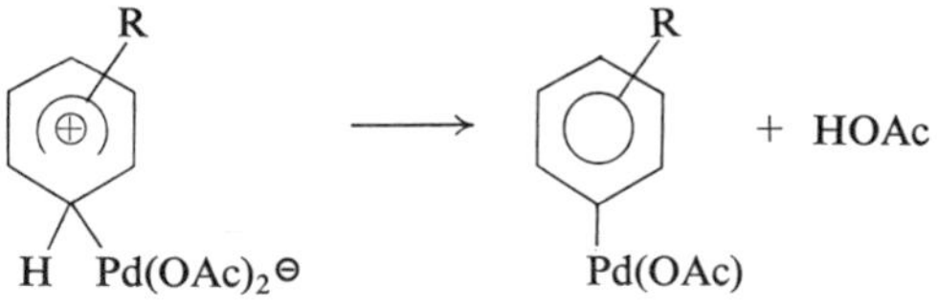

(I-54)

overall effects of the substituent R are then determined by the stabilities of the π complexes **(I-53)** rather than the σ complexes **(I-54)**. It is known from other studies that π-complex stability is much less influenced by substituent than is the corresponding σ complex.[243] Similar results have been obtained for the nitration of aromatic hydrocarbons by nitronium tetrafluoroborate in tetramethylene sulfone,[244] but in the palladation reactions ortho substitution is very heavily deactivated. This is presumably a steric effect.

Additional evidence for the intermediacy of a phenylpalladium complex comes from an observation by Tsuji and co-workers, quoted by Danno *et al.*[240], that the complex **(I-55)** was able to arylate styrene.

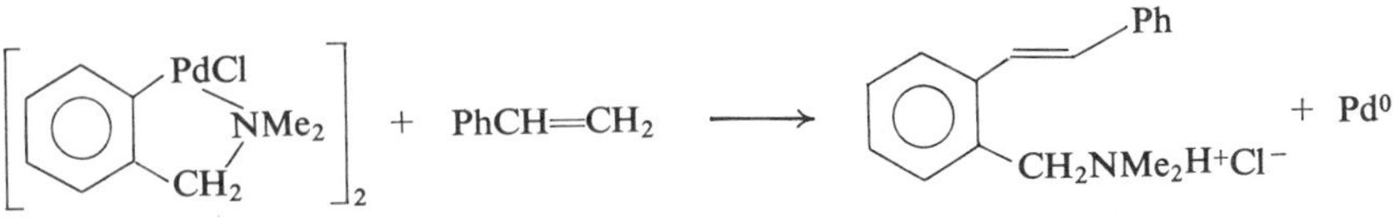

$$[\text{C}_6\text{H}_4(\text{PdCl})\text{CH}_2\text{NMe}_2]_2 + \text{PhCH=CH}_2 \longrightarrow \text{C}_6\text{H}_4(\text{CH=CHPh})\text{CH}_2\text{NMe}_2\text{H}^+\text{Cl}^- + \text{Pd}^0$$

(I-55)

The trans orientation of the product olefin is probably due to the fact that this is the thermodynamically favored isomer; in some cases, when the olefin is acrylonitrile or methyl acrylate, small amounts of the cis isomer are also obtained.[236, 240] Other simpler olefins, including ethylene, propylene, butenes, and butadiene also undergo this reaction, for example,[238]

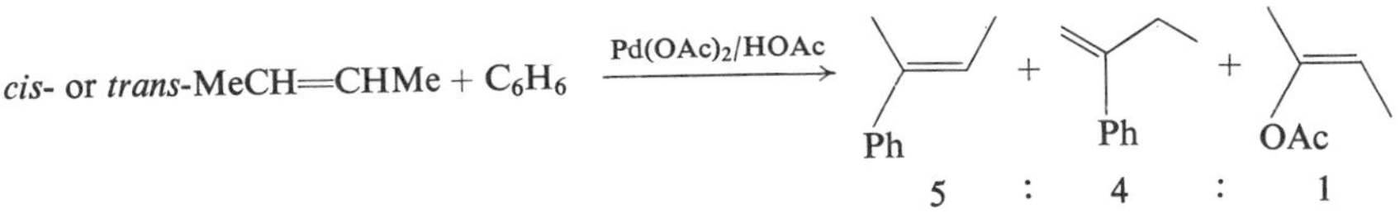

cis- or *trans*-MeCH═CHMe + C_6H_6 $\xrightarrow{Pd(OAc)_2/HOAc}$ (Ph-substituted butene) + (Ph-substituted butene) + (OAc-substituted butene)

5 : 4 : 1

However, Moritani *et al.* have suggested σ-vinyl complexes as intermediates in these reactions and have reported evidence in favor of this proposal (see Appendix).

E. MISCELLANEOUS REACTIONS LEADING TO C–C BOND FORMATION

1. Electrophilic Substitution

Smutny *et al.*[137] have noted that (1-chloromethyl)allylpalladium chloride dimer alkylated phenol, in the absence of catalyst and solvent at 55°, largely in the para position to give **(I-56)**. Reactions of this type are obviously

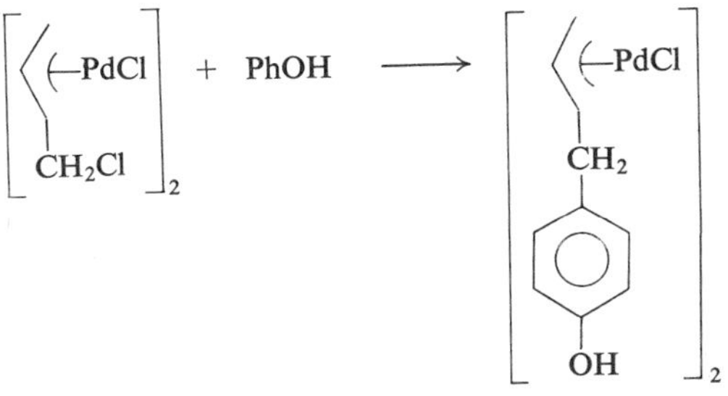

$$[(\text{C}_3\text{H}_4\text{CH}_2\text{Cl})\text{PdCl}]_2 + \text{PhOH} \longrightarrow [(\text{C}_3\text{H}_4\text{CH}_2\text{C}_6\text{H}_4\text{OH})\text{PdCl}]_2$$

(I-56)

$$[(\text{C}_3\text{H}_4\text{CH}_2^{\oplus})\text{PdCl}]$$

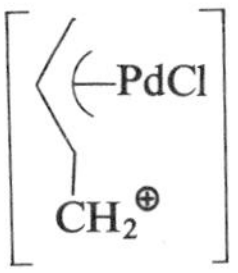

(I-57)

capable of further extension and must be due to the stability of the carbonium ion (**I-57**) or similar species (see also Volume I, Chapter V, Section F,3,b).

Palladium chloride can also act as a Friedel–Crafts catalyst in certain cases. For example, Long and Marples in a patent[245] describe the formation of mono-, di-, and triacid chlorides from the reaction of benzene with phosgene and CO in the presence of $PdCl_2$ in hexane.

$$C_6H_6 + COCl_2 + PdCl_2 \xrightarrow{CO/60\ atm/240^\circ} \underset{48\%}{C_6H_5COCl} + \underset{1\text{–}2\%}{C_6H_4(COCl)_2} + \underset{27\%}{C_6H_3(COCl)_3}$$

2. Formation of Propylene from Ethylene

Two remarkable papers, one by Crano *et al.*[110] and the other by Aguilo and Stautzenberger,[109] have described the formation of propylene (in about 15% yield) from ethylene, but under very different conditions. The former workers found that this reaction occurred when ethylene and $PdCl_2$ were heated in a weakly coordinating solvent such as benzonitrile in the presence of sodium fluoride at 160°. Other fluorides were less effective. In the latter case, reaction occurred at atmospheric pressure in the presence of HCl, acetic acid, and acetic anhydride. In both cases metal and some butenes were formed. The mechanism of reaction is not at all clear, but one possibility is that ethylene and 2-butene disproportionate under these conditions, to give propylene.

$$C_2H_4 + MeCH{=}CHMe \rightarrow 2MeCH{=}CH_2$$

This type of reaction, termed "olefin metathesis," is well known for other metal catalysts, particularly tungsten.[246] The 2-butenes, of course, arise by dimerization of ethylene (Section C,1).

3. Catalysis of Addition of Ethyl Diazoacetate to Olefins

As has been mentioned in Volume I, Chapter V, Section G, allylpalladium chloride dimer is a very effective catalyst for the reaction of ethyl diazoacetate with olefins to give cyclopropanes.[247]

F. CLEAVAGE OF CARBON–CARBON BONDS

A number of reactions are also known in which a (presumably) palladium-induced C–C bond cleavage occurs. Most of these have already been mentioned elsewhere and they are merely summarized here. Considerably fewer are known than those in which C–C bond formation occurs, but this may be due to lack of interest in these reactions.

They can be divided as follows: (1) reactions in which a complex is formed, (2) homogeneous reactions in which a stable complex is not formed, (3) heterogeneous reactions, and (4) skeletal rearrangements.

1. Formation of a Complex with C–C Bond Cleavage

The most interesting case is the ring-opening of cyclopropanes to π-allylic complexes (Volume I, Chapter V, Section B,3). If it is assumed that the π character of the cyclopropane[248] allows the transient formation of a complex in which there is donation of electron density from cyclopropane bonding orbitals to vacant palladium orbitals and back-donation from the metal into antibonding orbitals, then the net effect will be to weaken the C–C bonds and to cause the cyclopropane ring to open.

Hüttel and Schmid[249] have reported the formation of π-allylic complexes from allylic carboxylic acids with decarboxylation.

$$CH_2{=}C(Me)\text{—}C(Me)(X)COOH \xrightarrow{PdCl_2/HOAc} [Me\text{—}(\pi\text{-allyl})PdCl(Me)(X)]_2 + CO_2 + HCl$$

[X = COOEt (63%), CN (78%), $CONH_2$ (95%), and COOH (97%)]

A strongly electron-withdrawing group, X, is necessary. However this is not a general reaction and sometimes decarboxylation does not occur, for example,

$$C_6H_9\text{—}CH(COOH)_2 \xrightarrow{PdCl_2/HOAc} [C_6H_{10}\text{—}(\pi\text{-allyl})PdCl(COOH)(COOH)]_2$$

Johnson *et al.*[250] have reported the acid cleavage of a C–C bond in a palladium complex:

$$[CH(COMe)_2\text{-dicyclopentadiene-}PdCl]_2 \xrightarrow{HCl} \text{dicyclopentadiene-}PdCl_2$$

No heterolytic cleavage of the Pd–C σ bond was found.

2. Palladium-Induced Cleavage of C–C Bonds under Homogeneous Conditions

A number of reactions of this type were observed by Smidt and co-workers[98, 251] in the course of their studies on the oxidation of substituted olefins in the Wacker process, for example,

$$RCH{=}CHCONH_2 \xrightarrow{PdCl_2/H_2O} RCOMe$$

$$RCH{=}CHCOOH \longrightarrow RCOMe$$

(R = Me, Et, Pr, Bu, Ph; yields 48–94%)

A decarboxylation occurs in both of these reactions. Olefins with other electron-withdrawing substituents reacted normally; for example, acrylonitrile gave acetyl cyanide[251] (see also p. 67).

A variant on this is the decarboxylation of higher straight-chain aliphatic acids in the presence of the sodium salt, the appropriate anhydride, and $PdCl_2$.[229]

$$RCH_2CH_2COOH + RCH_2CH_2COONa + (RCH_2CH_2CO)_2O + PdCl_2 \longrightarrow RCH{=}CH_2 + CO_2 + Pd.$$

3. Palladium-Induced Cleavage of C–C Bonds under Heterogeneous Conditions

Tsuji *et al.*[252, 253] have found that $PdCl_2$, which is reduced during the reaction to palladium black (the real catalyst), was very effective for the decarbonylation of acid halides. In some cases aryl or benzyl halides resulted:

$$RCOCl \xrightarrow{PdCl_2/200°} RCl + CO$$

$$(R = Ph, PhCH_2)$$

while for acyl chlorides with a hydrogen β to the carbonyl, HCl elimination also occurred to give isomeric mixtures of monoolefins.

$$C_9H_{19}COCl \xrightarrow{PdCl_2/200°} C_9H_{18} + CO + HCl$$

Aldehydes were similarly decarbonylated to mixtures of saturated and unsaturated hydrocarbons.[253]

$$C_9H_{19}CHO \xrightarrow{PdCl_2/200°} C_9H_{20} + C_9H_{18} + CO$$

Other workers have examined the palladium on carbon catalyzed decarbonylation of aldehydes in more detail.[254,255] Wilt and Abegg[255] concluded that this was not a radical process as little rearrangement occurred in the decarbonylation of $PhCMe_2CH_2CHO$ to $PhCMe_3$. Ohno and Tsuji[253] found $[Ph_3PPtCl_2]_2$ to be inactive for decarbonylation, but $(Ph_3P)_3RhCl$ and $(Ph_3P)_2RhCOCl$ were very effective catalysts.

Tsuji *et al.* have proposed a mechanism for these decarbonylation reactions.[252] This mechanism can also explain the catalyzed hydrogenation of acyl halides to aldehydes (Rosenmund reaction). σ-Bonded organopalladium compounds are postulated as intermediates.

$$RCH_2CH_2COCl + Pd$$

$$\updownarrow$$

$$RCH{=}CH_2 + Pd + HCl \rightleftharpoons RCH_2CH_2PdCl \underset{}{\overset{CO}{\rightleftharpoons}} RCH_2CH_2COPdCl$$

$$\downarrow H_2\ (-HCl)$$

$$RCH{=}CH_2 + Pd + H_2 \rightleftharpoons RCH_2CH_2PdH \overset{CO}{\rightleftharpoons} RCH_2CH_2COPdH \rightleftharpoons RCH_2CH_2CHO + Pd$$

$$\downarrow$$

$$RCH_2CH_3$$

The reversibility, to some extent at least, of these reactions was suggested by the formation (in low yield) of propionaldehyde from ethylene, CO, and hydrogen catalyzed by Pd in an autoclave. However, other metals may have been responsible for this hydroformylation since Tsuji *et al.* reported that in a glass liner the reaction gave largely ethane.[74]

4. Palladium-Induced Skeletal Rearrangements

A number of reactions have recently been reported in which a rearrangement of the carbon skeleton of an organic molecule has occurred. In two cases, metal complexes of the rearranged product were isolated and characterized.

Trebellas *et al.*[256] found that *cis-trans*-1,5-cyclodecadiene underwent a palladium chloride-catalyzed isomerization to give a palladium chloride complex of *cis*-1,2-divinylcyclohexane (**I-58**).

$$+ \ (PhCN)_2PdCl_2 \longrightarrow$$

$PdCl_2$

(I-58)

4-Vinylcyclohexene has been reported[257] to give 1,5-cyclooctadienepalladium chloride (Volume I, Chapter IV, Section D,2).

Vedejs has reported the reaction of bullvalene with $(PhCN)_2PdCl_2$ to give a solid which on decomposition with pyridine gave the bicyclo[4.2.2]decatetraene **(I-59)**.[258]

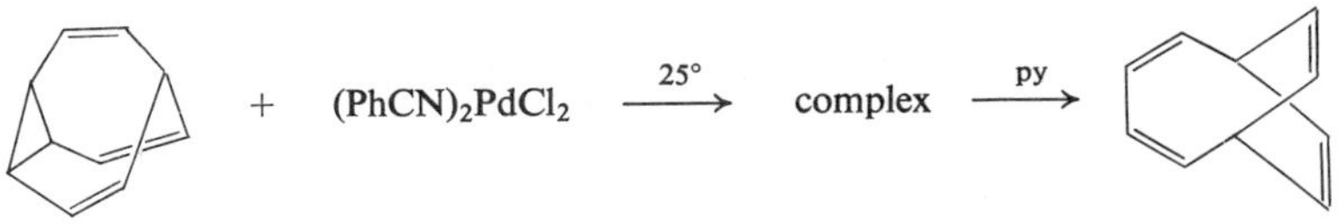

$$+\ (PhCN)_2PdCl_2 \xrightarrow{25°} \text{complex} \xrightarrow{py}$$

(I-59)

The normally forbidden disrotatory electrocyclic ring-opening of a Dewar benzene to the benzene has been shown to be catalyzed by a number of metal complexes, including $(PhCN)_2PdCl_2$. Dietl and Maitlis[259] showed that hexamethyl(Dewar benzene)palladium chloride **(I-60)** was converted to hexamethylbenzene by excess palladium chloride in solution. Kaiser and Maitlis[191]

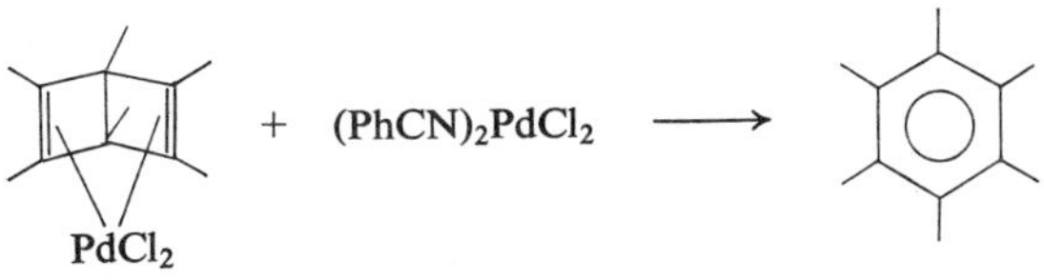

$$+\ (PhCN)_2PdCl_2 \longrightarrow$$

$PdCl_2$

(I-60)

have also shown that 1,2,5-tri-*t*-butyl(Dewar benzene) was isomerized by $(PhCN)_2PdCl_2$. In this case no complex was detected, and the products were a mixture of the 1,2,4- and the 1,3,5-tri-*t*-butylbenzenes in the ratio of 3:1. The isomerization was also effected thermally and with acid; in both cases no significant amounts of the 1,3,5-isomer were found.

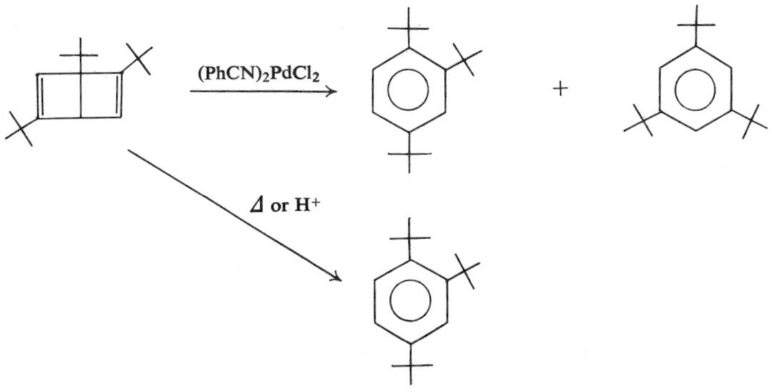

$$\xrightarrow{(PhCN)_2PdCl_2} \quad + \qquad \xrightarrow{\Delta \text{ or } H^+}$$

Palladium chloride (or silver perchlorate) also isomerizes cubane to cuneane.[260]

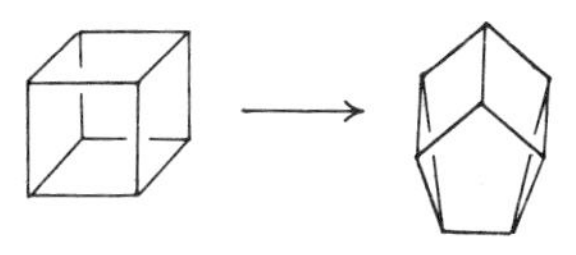

At first sight, these reactions appear to be examples of the metal catalysis of electrocyclic reactions which are thermally symmetry-restricted. The general subject of the metal catalysis of symmetry-restricted reactions has been extensively discussed. Explanations which have been offered include, (1) that the metal can use its d orbitals to lift the symmetry-imposed restrictions[261–263]; (2) that the metal can oxidatively add to the hydrocarbon to give a σ-bonded organometallic intermediate which then decomposes[264]; and (3) that for strained hydrocarbons the reaction proceeds via a charge-transfer complex.[265] It also appears probable, in view of the variety of such reactions which have been found, that several different mechanisms can occur. For example, a simple disrotatory ring opening of 1,2,5-tri-*t*-butyl (Dewar benzene) could give none of the 1,3,5-tri-*t*-butylbenzene found in the metal-catalysed reaction. In this case *at least*, a stepwise and nonconcerted mechanism would appear to be operative.

Chapter II

The Formation of Carbon–Oxygen Bonds

A. INTRODUCTION

Palladium(II) is very effective at inducing the formation of C–O bonds, for example,

$$C_2H_4 \rightarrow CH_3CHO$$

$$C_2H_4 \rightarrow CH_2{=}CHOAc$$

Since these reactions involve oxidation of the organic substrate and hence reduction of the Pd(II) (to metal), they are not catalytic. They can be made so, however, and hence become industrially attractive, by reoxidizing the palladium metal back to Pd(II), usually *in situ*. Any oxidizing agent which has a redox potential higher than that of Pd(II) can be used. In practice Cu(II) or Fe(III) are utilized commercially, as pioneered by Smidt and his collaborators in the Wacker process; *p*-benzoquinone has also been widely used, especially in mechanistic investigations. The use of Cu(II) or Fe(III) is particularly advantageous since *their* reduction products [Cu(I) and Fe(II), respectively] can be reoxidized by oxygen (or air). This has allowed the cyclic catalytic process:

$$\begin{array}{rcl}
C_2H_4 + PdCl_2 + H_2O & \rightarrow & CH_3CHO + 2HCl + Pd \\
2CuCl_2 + Pd & \rightarrow & 2CuCl + PdCl_2 \\
2CuCl + 2HCl + \tfrac{1}{2}O_2 & \rightarrow & 2CuCl_2 + H_2O \\
\hline
C_2H_4 + \tfrac{1}{2}O_2 & \rightarrow & CH_3CHO
\end{array}$$

to be developed.

This reaction and its numerous variations as well as related ones form the subject of this chapter. Other reactions involving C–O bond formation, but excluding some "trivial" ones (such as the solvolysis of acyl halides) which may also be catalyzed by Pd(II) but where little mechanistic information is available, are also reviewed.

B. THE OXIDATION OF OLEFINS

The basic principle of the reactions to be discussed is illustrated by the ease with which diene–$PdCl_2$ complexes undergo nucleophilic attack by alkoxide (Volume I, Chapter IV, Section F,1,c). This is in sharp contrast to uncomplexed olefins, which undergo electrophilic rather than nucleophilic attack.[266]

Chatt *et al.* in 1957 noted that (1,5-cyclooctadiene) palladium chloride and similar diene complexes reacted with alkoxide, particularly OMe^- or OEt^-, to give the alkoxycyclooctenyl complex (**II-1**) in which a C–O and a C–Pd σ bond had been formed.[267]

$PdCl_2$ $\xrightarrow{OR^-}$ [OR PdCl]$_2$

(**II-1**)

Diene–platinum(II) complexes also undergo this reaction,[268] but less easily; the products are also less reactive than the palladium complexes and this can be correlated with the general higher reactivity of palladium compared to platinum.

Numerous other workers have extended these reactions to other systems and have shown that only one product is obtained, this being the isomer with the alkoxy group exo to the metal.

This result strongly suggests that attack by the alkoxide is directly at a diene carbon and does not involve prior formation of an alkoxy–metal intermediate. If this assumption is correct, these reactions differ in detailed mechanism from those considered for monoolefins (below), where attack by coordinated alkoxide or hydroxide on the coordinated olefin is *usually* preferred to explain the kinetics. Reactions of either type are now commonly termed "oxy-palladation" reactions.

Panunzi *et al.*[269] have recently shown that attack by amines on monoolefin-Pt(II) complexes is exo. If this result can be applied to other systems and to

Pd(II), it suggests that for monoolefins the mode of attack depends on the precise conditions of the reaction (for example, the concentration of the nucleophile) as well as on the nature of the attacking ligand and its affinity for the metal (see Appendix).

1. Ethylene: Scope of the Reactions

Ethylene has been converted to a wide range of products in palladium-induced reactions under oxidizing conditions. The main products (few of which are obtained entirely free of others) are illustrated in Scheme II-1.

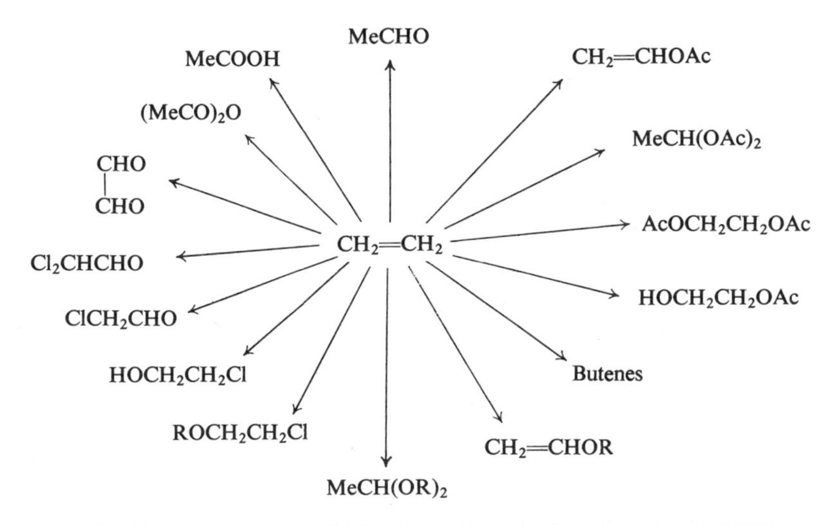

SCHEME II-1. Organic compounds obtained from the oxidation of ethylene by Pd(II).

In addition, a patent has reported formation of vinyl chloride (in the presence of $CuCl_2$)[270]; and these reactions can also be combined with carbonylation, for example, to give $AcOCH_2CH_2COOH$.[11]

Acetaldehyde (and acetic acid to some extent) and vinyl acetate are now largely prepared from ethylene by these reactions. They are favored over the older routes using acetylene as starting material, since ethylene is (currently) cheaper and presents fewer hazards in commercial operation. The synthesis of acetaldehyde from ethylene is commonly known as the Wacker process after the company where it was developed by J. Smidt and his collaborators in 1956. The patents for this process are mainly held by Consortium für Elektrochemische Industrie G.m.b.H. (a subsidiary of Wacker Chemie G.m.b.H.) and Farbwerke Hoechst A.G., in conjunction with which it was developed. Details of the patents have been listed by Szonyi[271]; pertinent ones and some of the more recent are also included here.

These reactions are also of interest in being among the few homogeneously catalyzed reactions which are already commercially attractive. In large part this is due to the volatility of the products, which makes separation and hence catalyst recovery fairly easy. Heterogeneous catalytic processes based on supported palladium catalysts and a gas-phase reaction have also been developed, but do not appear to be favored for the acetaldehyde reaction. In the vinyl acetate process a solid-state catalyst has been developed which is attractive if a cheap source of acetic acid is available.

Krekeler and Schmitz[272] have pointed out that in the liquid-phase vinyl acetate synthesis, acetaldehyde is always a by-product owing to the unavoidable presence of water in the catalytic reaction,

$$C_2H_4 + HOAc + \tfrac{1}{2}O_2 \rightarrow CH_2{=}CHOAc + H_2O$$

Since acetaldehyde is readily oxidized to acetic acid, this means that both starting materials are readily available here. In cases where acetic acid is regenerated, for example, after polymerization of vinyl acetate and hydrolysis to polyvinyl alcohol, no extra external source is necessary and the gas-phase heterogeneous process is more advantageous. This latter process is widely used today and is very similar to the homogeneous reaction. A study of it has been reported by Nakamura and Yasui[273] who also quote details of some relevant patents.

Patents covering the formation of glycol mono- and diacetates,† vinyl ethers,[276] and acetals[277, 278] from ethylene have also been taken out. It is not clear whether these materials are commercially produced by these reactions yet.

Under aqueous conditions the invariable product of the oxidation of ethylene is acetaldehyde. One patent[279] claims that in the presence of KCl (and $CuCl_2$) at a pH below 2.5–3.0 some 2-chloroethanol is also formed. Acetic acid can be obtained if the oxidation is run in the presence of cobalt(III) (or manganese) acetate[280–282]; the gas-phase reaction over a Pd–Au catalyst gives acetic anhydride and acetic acid in addition to acetaldehyde.[283] Under aqueous conditions in the presence of nitrate, glyoxal (OCHCHO) is formed in addition to acetaldehyde according to an ICI patent.[284]

The situation in nonaqueous media is much more complex and the products obtained depend very much on the amount and nature of the anions present. The copper(II) salts which are usually present to reoxidize the palladium also

† However, the original report[274] that ethylene glycol can be obtained from these reactions now appears to be incorrect.[275] Ethylene glycol and its monoacetate are very difficult to distinguish by v.p.c., and this probably caused the confusion.

appear to play an important, if rather ill-defined, role in some of these reactions. A further complication is that a number of reactions have been described using palladium acetate which contained nitrogenous impurities (these may be present as cyanides, see Chapter IV, Section B) arising from its mode of preparation.[285] It appears, in some cases at least, that these impurities exert a profound influence on the course of the reaction, particularly when reactions are carried out in chloride-free media.

From a survey by van Helden *et al.*[285] and from the patent literature, a picture of the factors determining the selectivity of the reactions in acetic acid, alcohols, and similar media has begun to emerge.

In acetic acid ethylene reacts with palladium acetate to give vinyl acetate, acetaldehyde, ethylidene diacetate [$MeCH(OAc)_2$], and butenes (and, in some cases, also products derived from the butenes). Acetaldehyde can arise in two ways; one is by a Wacker reaction since water is formed during the oxidative regeneration of the catalyst. The other way is by the palladium(II)-catalyzed reaction of vinyl acetate and acetic acid,[286–288]

$$CH_2{=}CHOAc + HOAc \rightarrow MeCHO + Ac_2O$$

Thus both under anhydrous and aqueous conditions, acetaldehyde is always a by-product in the liquid-phase reaction. In the gas phase, however, it can be minimized.[289]

The formation of ethylidene diacetate is favored by low concentrations of water and high temperatures and is the major product under some conditions [e.g., $Pd(OAc)_2/Fe(OAc)_3/LiCl/HNO_3/HOAc/90°–95°/O_2$ (1 atm)[285]]. Other workers found that on addition of weak bases such as DMF[290] or DMAC,[291] or if the reaction is carried out at higher pressures,[292] the amount of ethylidene diacetate formed decreases. van Helden *et al.*[285] suggested that this product arose from vinyl acetate by addition of acetic acid, but other workers found it to be formed at the same time as vinyl acetate.[293] Furthermore, Moiseev and Vargaftik[294] found that ethylidene diacetate formed in the presence of AcOD contains nearly no deuterium. It therefore appears not to arise from vinyl acetate to any significant extent.

Under the conditions mentioned above, significant by-products are 1,2-diacetoxyethane and 2-acetoxyethanol. The latter becomes the major product at higher nitrate ion concentrations.[288, 295–299]

Henry[300] found that 1,2-diacetoxyethane and 2-acetoxyethanol arise only (in the absence of nitrate) when copper is present as a reoxidant. The latter product is formed in the presence of water, whereas 1-acetoxy-2-chloroethane, together with 1,2-diacetoxyethane, is formed at high [$CuCl_2$] concentrations.[301, 302]

The formation of butenes, by dimerization of ethylene, has already been discussed in Chapter I, Section C,1. It is found to be an important side-reaction under more anhydrous conditions, especially at high ethylene pressures and high chloride ion concentrations.

Since fairly high chloride ion concentrations are needed to keep the catalyst active, tricky problems arise in the commercial utilization of these processes. However, it appears that the reactions *can* be controlled to give any one of the above materials as major product.[302]

Much less intensive work has been done on the reactions of alcohols with olefins in the presence of palladium chloride. This reaction, like that with acetic acid, is strongly catalyzed by base; the major products at low temperatures are the acetals, while at higher temperatures vinyl ethers are formed.[102,277,278,303–306]

Copper(II), usually as the chloride or acetate, is the most widely used reoxidant although *p*-benzoquinone has also been used.[301,307] Base, usually as acetate, and chloride (preferably as the lithium salt owing to its greater solubility) are also added to the reaction mixture. Certain other additives, such as Fe(III) acetate, nitrate, rare-earth salts,[308] DMF, and DMAC, are often beneficial (see below).

2. Oxidation of Ethylene to Acetaldehyde in Water

Phillips in 1894 first noted that aqueous palladium chloride was reduced to metal by ethylene and that acetaldehyde was formed.[157] Anderson in 1934 observed that Zeise's salt, $K[PtC_2H_4Cl_3]$, reacted similarly with water on heating.[309] However, this reaction was much less efficient than that with palladium (ca. 60% of the ethylene was liberated unchanged) and needed more drastic conditions. It was not until the work of Smidt and co-workers, who found that the palladium metal formed in the reaction could easily be reoxidized, that the synthetic possibilities of this reaction were recognized.

In the years since the work of Smidt considerable attention has been paid to the mechanisms of the reaction by a large number of workers including Smidt, Moiseev, Henry, and their collaborators. Work in this field has been reviewed by Aguilo,[310] Henry,[311] Stern,[312] and Moiseev.[313] The account given here is based on these reviews, together with more recent references.

As already mentioned, the success of the Wacker process depends on the presence of an oxidizing agent to reoxidize the palladium up to the (II) state. Copper(II) salts are usually used since the reoxidation of Cu(I) to Cu(II) by oxygen is fast. However, the reoxidation step [Cu(I) → Cu(II)] is overall the slow step and is rate-determining for the catalytic reaction.[314] In practice, the reaction can be run as a single stage process, in which ethylene and oxygen are passed through the catalyst solution, or as a two-stage process. In the latter, ethylene and some air (or oxygen) are first fed into the catalyst solution until all the Cu(II) has been converted to Cu(I). The catalyst is then pumped into a separate vessel where it is reoxidized by air.[271]

The work of Anderson on Zeise's salt and subsequent work on the much more reactive (olefin)palladium(II) complexes suggested, but did not prove, that olefinic π complexes are intermediate in these reactions. Since (olefin)-palladium(II) complexes are relatively unstable in water, under the conditions of the Wacker process, they cannot be isolated and evidence for their participation is indirect.

Under appropriate conditions, where further reaction was slow, Moiseev *et al.*[315,316] and Henry[317,318] showed that ethylene (and other olefins) dissolved in solutions of Pd(II) salts to a greater degree than in the absence of palladium. This observation was developed into a method for determining the stability constants, but the precise physical interpretation of the data is still the subject of some dispute.[313]

Both Moiseev and Henry agreed that only 1:1 olefin:palladium complexes were formed under their conditions, and that two equilibria (K_1 and K_2),†

$$\text{olefin} + PdCl_4^{2-} \xrightarrow{K_1} (\text{olefin})PdCl_3^- + Cl^-$$

$$(\text{olefin})PdCl_3^- + H_2O \xrightarrow{K_2} (\text{olefin})PdCl_2(H_2O) + Cl^-$$

exist. These conclusions arose from the lack of any effect of change of the $[H^+]$ concentration over a large range and the inverse effect of change of chloride ion concentration. Values for K_1 are quoted in Volume I, Chapter III, Section C; K_2 was estimated at ca. 10^{-3} *M* by Henry,[318] but much higher by Pestrikov *et al.*,[316] who postulated that the concentrations of $C_2H_4PdCl_3^-$ and of $C_2H_4PdCl_2(H_2O)$ were of the same order of magnitude.[313]

A number of possible mechanisms for the formation of acetaldehyde from the ethylene π complex have been considered over the years. Some of these can be ruled out. For example, Hafner *et al.*[319] and Nikiforova *et al.*[320] showed that ethanol is oxidized by $PdCl_2$ to acetaldehyde much more slowly than is ethylene. Ethanol cannot be an intermediate in the ethylene oxidation, e.g.,

$$H_2O + C_2H_4 \xrightarrow{Pd(II)} C_2H_5OH \xrightarrow{Pd(II)} MeCHO$$

since if it were, it would accumulate and be easily detectable. Similarly Krekeler (quoted by Hafner *et al.*[319]) noted that acetaldehyde formed from

† At the $[Cl^-]$ concentrations used, the palladium(II) was largely present as $[PdCl_4]^{2-}$.[310]

ethylene oxidized by Pd(II) in D_2O did not contain any deuterium. This disposed of an early mechanistic proposal that vinyl alcohol was an intermediate,[303] since in this case one hydrogen on the methyl must arise from $OH^-(OD^-)$.

$$C_2H_4 + OH^- + Pd(II) \rightarrow CH_2{=}CHOH + H^+ + Pd(0)$$

$$CH_2{=}CHOH \rightarrow CH_3CHO$$

Smidt *et al.* in their original study of the reaction[98, 319] found that the rate of oxidation was inhibited by H^+ and chloride ions. Bromide and iodide had an even greater effect, the relative rates decreasing in the order, F^- (11) > Cl^- (6) > Br^- (3.5) > I^- (1). In order to minimize complications arising from the presence of copper ions, the reactions were usually run catalytically in the presence of *p*-benzoquinone as reoxidant.[313, 321] The reactions were followed by rate of ethylene uptake[317] and by changes in the redox potential of the quinone–hydroquinone couple.[313, 322] The latter was shown to be essentially independent of Pd(II) and olefin concentrations.

The rate was found by all the workers to[313, 317, 321] be approximately first-order with respect to ethylene and $PdCl_4^{2-}$ concentration, and to vary as the inverse of $[H^+]^1$ and $[Cl^-]^2$,

$$\text{Rate} = \frac{-d(C_2H_4)}{dt} = \frac{k[PdCl_4^{2-}][C_2H_4]}{[H^+][Cl^-]^2}$$

The chloride inhibition was consistent with:

$$C_2H_4 + PdCl_4^{2-} \rightleftharpoons C_2H_4PdCl_3^- + Cl^-$$

$$C_2H_4PdCl_3^- + H_2O \rightleftharpoons C_2H_4PdCl_2(H_2O)^{\dagger} + Cl^-$$

and the acid inhibition suggested that the reaction involved nucleophilic attack on the olefin by OH^-. This could be either (A) by direct attack by uncoordinated OH^- or (B) by "insertion" of the olefin into the Pd–OH bond.

$$OH^- \curvearrowright \begin{matrix} CH_2 \\ \| {-}PdCl_2OH_2 \\ CH_2 \end{matrix} \rightarrow [HOCH_2CH_2PdCl_2 \cdot H_2O] \rightarrow \text{products} \qquad (A)$$

† Ethylene is a strongly trans-directing ligand in substitution reactions (Volume I, Chapter I, Section B,4,e), hence the initial product would be expected to be *trans*-$C_2H_4PdCl_2(H_2O)$. However, since Pd(II) complexes are very labile this complex may well exist largely as the more polar cis isomer in a polar solvent such as water.

$$C_2H_4PdCl_2(H_2O)^\dagger \rightleftharpoons H^+ + \left[\begin{array}{c} CH_2{=}CH_2 \to Pd(Cl)_2(OH) \end{array}\right]^-$$

(II-2)

$$\left[CH_2{=}CH_2 \to PdCl_2(OH)\right]^- \xrightarrow{L} [HOCH_2CH_2PdCl_2(L)]^- \longrightarrow \text{products} \quad \text{(B)}$$

($L = Cl^-$ or solvent)

Henry[317] has presented an objection to the reaction path (A) in that the rate constant of this reaction, if it occurred, would have to be ca. 10^{13} M^{-1} sec^{-1}, about 10^4 larger than the rate for a diffusion-controlled process in solution. This objection did not arise for path (B) and he suggested that reaction occurred mainly in this way.

Moiseev *et al.*[323] also agreed with this conclusion on the basis of studies in D_2O and H_2O. They found that the ratio of rates $k_{(H_2O)}/k_{(D_2O)}$ for the overall reaction was 4.05 ± 0.15 and argued that, assuming the formation constant of the π-ethylene complex in D_2O was the same as in H_2O, and that the nucleophilicities of OH^- and OD^- were the same, then $k_{(H_2O)}/k_{(D_2O)}$ should be equal to the ratio of the autoprotolysis constants for H_2O and D_2O, if attack occurred by free OH^- and OD^-. This was 5.08 under their conditions, and sufficiently different to the value of 4.05 observed that they ruled out direct attack by a nucleophile on the coordinated olefin.

Again, two possible routes for decomposition of the π complex **(II-2)** to products (acetaldehyde, HCl, and Pd) can be envisaged. One is a concerted reaction path, (C), and the other, (D), involves a cis-insertion reaction

$$\left[CH_2{=}CH(H) \to PdCl_2(O{-}H)\right]^- \longrightarrow CH_3CHO + HPdCl_2^- \quad \text{(C)}$$

(II-2)

(migration of coordinated OH) to give the σ-bonded β-hydroxyethylpalladium complex intermediate, **(II-3)**, which then undergoes a 1,2-hydride shift.

$$\left[CH_2{=}CH_2 \to PdCl_2(OH)\right]^- \longrightarrow \left[H{-}CH(O{-}H){-}CH_2{-}PdCl_2\right]^- \longrightarrow CH_3CHO + PdCl_2^{2-} + H^+ \quad \text{(D)}$$

(II-2) (II-3)

If (**II-2**) decomposed directly (path C) then a large isotope effect would be expected on replacing C_2H_4 by C_2D_4. On the other hand, if the σ complex (**II-3**) (path D) was formed first in a rate-determining step, then the observed isotope effect, k_H/k_D, would be small as the hydride transfer occurred in a fast step. In fact, Henry[317] found k_H/k_D to be 1.07, and therefore in better agreement with path (D). A direct demonstration of this point was provided by Moiseev and Vargaftik[324] who showed that 2-hydroxyethylmercuric chloride reacted with palladium chloride in ether to give acetaldehyde.

$$HOCH_2CH_2HgCl + PdCl_2 \xrightarrow{Et_2O} Pd + Hg + HgCl_2 + CH_3CHO$$

The reaction presumably went via a "$HOCH_2CH_2PdX$" intermediate. Other possible routes, for example, via an ethylene complex, were excluded. Similarly, $ROCH_2CH_2HgCl$ reacted with $PdCl_2$, to give $ROCH{=}CH_2$ (R = Et, Ac).

Henry[325] has also recently shown that

$$k_H'/k_D' = \text{H shift/D shift} = \frac{[CH_2DCDO]}{[CHD_2CHO]}$$

had a value of at least 1.5 for the reaction:

$$HDC{=}CDH \longrightarrow [Cl_3PdCHDCHDOH]^{2-} \begin{cases} \xrightarrow{\text{H shift}} CH_2DCDO \\ \xrightarrow{\text{D shift}} CHD_2CHO \end{cases}$$

Hence there does appear to be an appreciable isotope effect in these reactions. However, there is little data on isotope effects on hydride transfer processes of this type and therefore analogies are hard to find.

At high chloride ion and low palladium(II) concentrations, that is, under the usual conditions of the Wacker process, except that copper is not present, the above picture is able to account for the facts reasonably. This consists, in summary, of:

(1) Formation of $[C_2H_4PdCl_3]^-$ from ethylene and $PdCl_4^{2-}$ (reversible).

(2) Replacement of coordinated Cl^-, probably trans to C_2H_4, by H_2O, and a fast isomerization to give *cis*-$[C_2H_4PdCl_2 \cdot H_2O]$ (reversible).

(3) Acid dissociation of the aquo to a hydroxo complex, *cis*-$[C_2H_4PdCl_2OH]^-$ (reversible).

(4) Insertion of ethylene into Pd–OH to give a σ complex,† $[HOCH_2CH_2PdCl_2(L)]^-$ (probably irreversible).

† This cis insertion probably occurs only when the olefinic carbons, the metal, and the oxygen are coplanar.[326] This can be achieved if the olefin can rotate about the metal–olefin axis from its normal perpendicular orientation to a coplanar one. There is considerable evidence that this can and does easily occur (Volume I, Chapter III, Section C).

(5) Decomposition via a hydride shift in the σ complex to acetaldehyde (irreversible and very fast). This step is discussed a little more fully in Chapter III.

This scheme may not be completely correct at low chloride ion concentrations, where Jira *et al.*[321] found rather different kinetics, probably due to the Pd(II) not being present as $PdCl_4^{2-}$ but as hydrolyzed and polymerized species.[311, 313] Moiseev *et al.* also reported deviations from the reaction scheme at high Pd(II) concentrations,[327] but this has been disputed by Henry.[325] The Russian workers suggested that the rate law observed,

$$-d(C_2H_4)/dt = k[PdCl_4^{2-}][C_2H_4]/[H^+][Cl^-]^2$$

is a limiting expression valid only at low palladium concentrations and in the presence of extra chloride ions. Under other conditions, Moiseev *et al.*[327] found that a two-term rate law applied.

$$\frac{-d(C_2H_4)}{dt} = \frac{k_I[PdCl_4^{2-}][C_2H_4]}{[H^+][Cl^-]^2} + \frac{k_{II}[PdCl_4^{2-}]^2[C_2H_4]}{[H^+][Cl^-]^3}$$

At 0.02 *M* $PdCl_4^{2-}$ the first term contributed 75%, and the second only 25%, whereas at 0.2 *M* $PdCl_4^{2-}$, the first term contributed 25% and the second, 75%. Moiseev *et al.* have proposed a reaction scheme rather similar to that discussed above to explain these results. The main difference lies in the suggested attack of $PdCl_4^{2-}$ on the π-olefin complex (**II-2**) to give the binuclear complex (**II-4**) which can then react via a binuclear σ complex to give acetaldehyde and a Pd(I) species $Pd_2Cl_4^{2-}$.

$$PdCl_4^{2-} + C_2H_4PdCl_2OH^- \rightleftharpoons \left[Cl_2Pd(\mu\text{-}Cl)_2PdCl(OH)(CH_2{=}CH_2) \right]^{2-} + Cl^-$$

(**II-2**) (**II-4**)

$$(\textbf{II-4}) \rightleftharpoons \left[Cl_2Pd(\mu\text{-}Cl)_2PdCl(CH_2CH_2OH) \right]^{2-} \longrightarrow Pd_2Cl_4^{2-} + CH_3CHO + HCl$$

An alternative explanation, that the initial reactive species was $Pd_2Cl_6^{2-}$, was rejected, since under their conditions the concentration of such species was negligibly small; but see Henry and Marks.[327a]

The reactions described here have some interesting analogies in the reactions of olefins with Tl(III) and Hg(II). The oxymercuration reaction has been reviewed by Kitching,[328] and Henry has given a review[311] of his work on the oxidation of olefins by Tl(III).[329] Unfortunately, direct comparisons are not possible since both $HgCl_2$ and $TlCl_3$ are inactive for these reactions. The active species usually used are the acetates (or the nitrates), which form, as they do for Pd(II), weaker and hence more reactive complexes. The real comparison should therefore be with palladium acetate, and here unfortunately too little work under sufficiently rigorous conditions has been done. Since, on the basis of such work as has appeared, palladium acetate reacts rather differently to the chloride, the comparisons which can be made at present with Tl(III) and Hg(II) are only of a very qualitative nature.

Although mercury(II) does oxidize olefins to organic products, these reactions are not well characterized. In contrast to palladium, however, stable σ bonded adducts are formed reversibly.

$$Hg(OAc)_2 + C_2H_4 \rightleftharpoons AcOHgCH_2CH_2OAc$$

For unstrained olefins the addition is stereospecifically trans and a mechanism such as the following has been suggested.

$$C_2H_4 + Hg(OAc)_2 + H^+ \longrightarrow \begin{matrix} CH_2 \\ \| \text{—} HgOAc^+ \\ CH_2 \end{matrix} + HOAc$$

$$Y^- + \begin{matrix} CH_2 \\ \| \text{—} HgOAc^+ \\ CH_2 \end{matrix} \longrightarrow Y\text{–}CH_2CH_2HgOAc$$

Bearing in mind the very important differences in ligand, Henry has shown that for oxidation of ethylene by Tl(III) (to acetaldehyde and ethylene glycol), the rate-determining step was attack by Tl(III) on the olefin; added perchlorate and methyl substitution of the olefin greatly increased the rate. These results were taken to imply that the properties of the intermediate here (e.g., $HOCH_2CH_2Tl^{2+}$) were different to those of the analogous palladium complex (**II-3**) in that heterolytic cleavage (to give the glycol) occurred more readily.

On present evidence, therefore, Tl(III) compounds, perhaps because the metal is more electropositive and in a higher oxidation state, follow path (E), heterolysis by base (H_2O or OH^-), while Pd(II) compounds react by path (F), in which a β hydrogen shift occurs.

$$HOCH_2CH_2\text{—}M^{n+} \xrightarrow{(E)} \underset{\uparrow \text{ base}}{HOCH_2\text{—}CH_2\text{—}M^{n+}} \longrightarrow HOCH_2CH_2OH + M^{(n-2)+} + \text{base } H^+$$

$$HOCH_2CH_2\text{—}M^{n+} \xrightarrow{(F)} \underset{\text{base}\ \ H}{HOCHCH_2\text{—}M^{n+}} \longrightarrow O{:}CHCH_3 + M^{(n-2)+} + \text{base } H^+$$

As Henry has pointed out, 1-chloro-2-propanol undergoes acid hydrolysis analogously, to give acetone and 1,2-propylene glycol.

$$MeCH\begin{matrix} OH \\ CH_2Cl \end{matrix} \longrightarrow MeCOMe + MeCHOHCH_2OH$$

Relatively little work has been carried out on the copper-catalyzed reaction under the conditions of the Wacker process, and some workers have assumed the Pd(II)-induced oxidation of ethylene and the reoxidation of Pd(0) [and Cu(I)] to occur by independent routes. Matveev *et al.*[330, 331] reported that the rate did increase with an increase in [Cu^{2+}], and suggested that a second pathway, involving a Pd(II)–Cu(II) complex was important at higher copper(II) concentration. However, Aguilo[310] has pointed out that under the conditions of this reaction addition of Cu(II) effectively reduced the chloride ion concentration. This would then accelerate the overall rate observed as discussed above.

A recent study by François[332] suggested that two reaction paths indeed occurred. One involved the participation of copper(II), while the other, at lower [Cu^{2+}] concentrations, did not and was identical to that discussed above. The first path involved two $CuCl_3^-$ anions and it was suggested that the reactive intermediate was $[C_2H_4PdCl_2(H_2O)(CuCl_3)_2]^{2-}$, with the $CuCl_3$ ligands coordinated directly to the palladium. This seems unlikely and a chloride-bridged species, for example,

$$\left[\begin{matrix} Cl_3Cu{-}Cl \\ Cl_3Cu{-}Cl \end{matrix} > Pd \begin{matrix} CH_2 \\ \| \\ CH_2 \\ OH_2 \end{matrix}\right]^{2-}$$

appears more reasonable. This could then break down in a similar fashion to that previously described, but *without* the formation of Pd(0).

$$\left[\begin{matrix} & Cu^{II}Cl_3 & \\ & | & \\ & Cl & \\ & | & CH_2 \\ Cl_3Cu^{II}{-}Cl{-} & Pd & \| \\ & | & CH_2 \\ & H{-}O & \end{matrix}\right]^{3-} \longrightarrow \left[\begin{matrix} Cu^{II}Cl_3 \\ | \\ Cl \\ | \\ Cl_3Cu^{II}{-}Cl{-}Pd{-}CH_2{-}CH{-}O{-}H \\ \quad\quad\quad\quad\quad\quad\quad | \\ \quad\quad\quad\quad\quad\quad\quad H \end{matrix}\right]^{3-}$$

$$\longrightarrow CH_3CHO + H^+ + PdCl_2 + 2Cu^{I}Cl_3^{2-}$$

Instead, the two electrons are passed via the palladium and the chlorine bridges, one onto each copper atom which is thus reduced to copper(I).

3. Oxidation of Higher Olefins to Ketones and Aldehydes in Water

The reaction with palladium chloride and water is a perfectly general one undergone by most olefins, in particular, α-olefins. In general, α-olefins give rise to methyl ketones,[98]

$$RCH{=}CH_2 \rightarrow RCOMe \qquad (R = Me, Et, Pr, \ldots C_8H_{17}, Ph, PhCH_2)$$

cyclic olefins up to cycloheptene give the cycloalkanones,[98,333]

$$(CH_2)_n\begin{array}{l} CH \\ \| \\ CH \end{array} \longrightarrow (CH_2)_n\begin{array}{l} CH_2 \\ | \\ C{=}O \end{array} \qquad (n = 3, 4, 5)$$

and dienes react with isomerization,[98]

$$CH_2{=}CHCH{=}CH_2 \rightarrow MeCH{=}CHCHO$$

$$CH_2{=}CHCH_2CH{=}CH_2 \rightarrow EtCH{=}CHCHO$$

Smidt and his co-workers[98,251,334] have also investigated a wide variety of substituted olefins. Typical reactions are

$$RCH{=}CHCOX \rightarrow RCOMe \qquad (R = \text{alkyl, phenyl}; X = OH, NH_2)$$

$$RCH{=}CHBr \rightarrow RCOMe \qquad (R = \text{alkyl})$$

$$RCCl{=}CH_2 \rightarrow RCOMe \qquad (R = Me, Ph, PhCH_2)$$

$$YCH{=}CH_2 \rightarrow YCH_2CHO \qquad (Y = NO_2, CN)$$

Since α-olefins usually gave rise to methyl ketones, using the mechanistic ideas developed above, Markovnikov attack on the olefin can be said to have occurred with OH^- attacking the β carbon. However, in many reactions, some of the aldehydes, presumably arising via anti-Markovnikov addition, were also found. Hafner *et al.*[319] investigated this reaction and found that the amount of aldehyde depended on the pH, the nature of the palladium complex, the temperature, and the olefin. For a series of α-olefins the following percentages of aldehydes were obtained in aqueous perchloric acid at 70°: propylene, 15%; 1-butene, 9%; 1-pentene, 20%; 1-hexene, 4%; 1-heptene, 5%; and styrene, 75%. Propylene was reacted under a variety of conditions and it was found that a maximum yield of aldehyde was obtained using K_2PdCl_4 in very strongly acid media (18 *M* $HClO_4$ or 10 *M* HCl) at 80°. The steric effect of the substituent on the olefin was thought to be of great importance. For example, in 1-pentene the alkyl group shields the β-olefinic carbon very effectively, giving rise to considerable attack at the α position.

Okada *et al.*[335, 336] have studied the relative amounts of acetophenones and phenylacetaldehydes obtained from various styrenes.

$$XC_6H_4CH{=}CH_2 \rightarrow XC_6H_4COMe + XC_6H_4CH_2CHO$$

They found that electron-withdrawing substituents, X, facilitated the formation of the aldehydes, whereas electron-releasing substituents, X, facilitated the formation of the ketones. The rates of formation of both products were first-order in both the styrene and the palladium complex, but showed a complex dependence on the chloride ion concentration. The measurements were, however, carried out at low chloride ion concentrations, and the remarks made on studies of the ethylene oxidation under these conditions apply here too.

It is assumed that both aldehydes and ketones arise from the same type of intermediate, i.e., (**II-5**). The effect of substituents in the phenyl ring could be to direct the insertion of the olefin into Pd–OH. Alternatively, direct attack by OH^- on the complexed olefin may be occurring.

$$\left[\begin{array}{c} CH_2 \\ \| \!\!-\!\! Pd(Cl)_2(OH) \\ CH \\ R \end{array} \right]^- \quad \nearrow \; [RCH(OH)CH_2PdCl_2(L)]^- \rightarrow RCOCH_3$$

$$(\textbf{II-5}) \quad \searrow \; [HOCH_2CH(R)PdCl_2(L)]^- \rightarrow RCH_2CHO$$

As mentioned above, olefins with strongly electron-withdrawing substituents (NO_2, CN) give the expected product, where attack has occurred at the β carbon.

In contrast, halides on either the α or the β carbon are removed during the reaction and the methyl ketones result. Olefins with β-carboxylic acid or amide groups behave similarly. Allyl chloride undergoes a two-stage reaction; first, to give acrolein under mild conditions, perhaps via the π-allyl complex, and then to give methylglyoxal.[98, 214, 334]

$$CH_2{=}CHCH_2Cl \rightarrow CH_2{=}CHCHO \rightarrow CH_3COCHO$$

Hüttel *et al.*[214, 333, 337] have shown that unsaturated aldehydes sometimes arise from oxidation of branched chain olefins, and have suggested that they are formed by oxidation of intermediate π-allylic complexes (Volume I, Chapter V, Section F,7 and Section I,1 in this chapter).

A further reaction sometimes observed is the cleavage of a C=C bond to give a ketone; this is particularly true for α-substituted styrenes,[213, 214]

$$\begin{matrix}Ph \\ R\end{matrix}\!\!>C{=}CH_2 \xrightarrow{PdCl_2/HOAc/H_2O} PhCOR + PhRC{=}CHCH{=}CPhR$$

(R = Me, Ph)

Suggestions have been made that in some cases the oxidation of olefins other than ethylene follows a path different to that described in the previous section. For example, Vargaftik *et al.*[322, 338] found that the rate of formation of cyclohexanone from cyclohexene, while first-order in olefin and palladium chloride, was independent of pH.

In general, however, for the lower olefins the same picture as for ethylene has emerged. For example, Henry[318] found that the oxidation of propylene to acetone, and butenes to methyl ethyl ketone followed the rate law:

$$\frac{-d(\text{olefin})}{dt} = \frac{k'K[\text{olefin}][PdCl_4{}^{2-}]}{[Cl^-]^2[H^+]}$$

where K was the equilibrium constant for π-complex formation in water. As both K and k' decreased with increasing chain length and substitution, this led to an overall decrease in rate for the higher substituted olefins.

The most interesting point which emerged from this study was the very small effect of changes of structure of the olefin on the rate constant, k'. The change was in the order: ethylene > *trans*-2-butene ≈ propylene > *cis*-2-butene ≈ 1-butene, but the total range was only a factor of 6. This contrasts with the oxymercuration and oxythallation reactions where a very large *increase* (160-fold) in rate was observed on going from ethylene to propylene or isobutene.[311, 329] These results led Henry to propose that in the insertion into Pd–OH, the rate-determining step, there was little carbonium ion character in the transition state. If there had been, as was the case for insertion into Tl–OAc, a very large increase in rate for the substituted ethylene should have been observed.

$$Cl_2Pd\cdots CH_2{\cdots}CHR\cdots O(H) \xrightarrow{(L)} Cl_2Pd(L)CH_2CHR(OH)$$

(L = solvent)

It would be of great interest to see whether this pattern was retained for reactions of palladium acetate under more comparable conditions.

Very similar conclusions were arrived at by Moiseev and his co-workers[316, 322, 327, 339] and Dozono and Shiba[340] on the basis of their work. However, in one paper[327] it was noted that for propylene and butenes (as for ethylene) the rate became second-order with respect to $[PdCl_4{}^{2-}]$ at high palladium and low chloride concentrations.

Patents have been taken out covering the production of acetone,[341–344] methyl ethyl ketone (MEK),[342, 345–348] and higher α-olefins to methyl ketones.[342–345, 349–353] The commercial synthesis of acetone and MEK by this route appears attractive, and both one- and two-stage processes have been proposed by Smidt.[354] Clement and Selwitz have described a convenient synthesis of methyl decyl ketone from 1-dodecene using aqueous DMF as solvent.[355]

An interesting extension of these reactions has recently been described by Ouellette and Levin[356] who reacted phenylcyclopropane with palladium chloride in water. The cyclopropane again showed some olefinic character and gave propiophenone (60%) and benzyl methyl ketone (35%).

$$\text{Ph}\!-\!\triangleleft + PdCl_2 + H_2O \rightarrow PhCOEt + PhCH_2COMe + PhCH{=}CHMe$$

Experiments in D_2O showed that no deuterium was incorporated into the products. Phenylcyclopropane was also isomerized to 1-phenyl-1-propene at low Pd(II) ratios and the benzyl methyl ketone arose from oxypalladation of this olefin. Very high ratios of palladium chloride to phenylcyclopropane gave a 95% yield of propiophenone (see Volume I, Chapter V, Section B,3 and this Volume, Chapter I, Section F,1). The formation of propiophenone can be explained in terms of addition of Pd–OH to the 1,2-bond of the cyclopropane followed by two consecutive 1,2-hydride shifts.

4. Oxidation of Ethylene in Acetic Acid: The Formation of Vinyl Acetate, Ethylidene Diacetate, and Glycol Esters

Moiseev *et al.* in 1960 reported that the ethylenepalladium chloride complex was stable in acetic acid,† but that on addition of acetate a rapid reaction occurred to give vinyl acetate and palladium metal.[303] The same reaction occurred with ethylene and $PdCl_2$, instead of the complex, and could be run catalytically in the presence of quinone as reoxidant.

Stern and Spector[304] independently reported the same reactions using HPO_4^{2-} as base in acetic acid. They also reported that propylene gave isopropenyl acetate and butadiene gave 1-acetoxybutadiene.

Although this area has been intensively studied during the years since these discoveries were announced, very little has until recently appeared in the open literature. A survey of the patents issued subsequently to cover this reaction, as well as the more recent literature, shows that vinyl acetate ($CH_2{=}CHOAc$) is not the only product from this reaction. Other products, the

† Some dimerization to butenes did occur.

amounts of which depend on the exact conditions, include acetaldehyde, ethylidene diacetate [$CH_3CH(OAc)_2$], 1,2-diacetoxyethane [$AcOCH_2CH_2OAc$], 2-acetoxyethanol [$AcOCH_2CH_2OH$], and 1-acetoxy-2-chloroethane [$AcOCH_2CH_2Cl$]. The latter three compounds arise only in the presence of nitrate[295, 300] or copper(II) salts. Acetaldehyde and ethylidene diacetate are formed under all conditions, preferentially so in the absence of Cu(II) and nitrate.

$$C_2H_4 + HOAc \longrightarrow CH_2{=}CHOAc + MeCHO + MeCH(OAc)_2 + AcOCH_2CH_2OAc + AcOCH_2CH_2OH + AcOCH_2CH_2Cl$$

In addition, acetic acid and acetic anhydride can be obtained from subsequent reactions (see also Szonyi[271]).

Moiseev and Vargaftik[294] showed that when the reaction was carried out in CH_3COOD, virtually no deuterium was incorporated into the above products.

In the quinone-catalyzed reaction Moiseev and his co-workers[357] and Ninomiya *et al.*[358] showed that the rate of ethylene absorption was independent of quinone concentration and first-order with respect to ethylene and Pd(II) concentration. Moiseev *et al.*[359,360] have also reported some more detailed studies on the reaction of ethylene with palladium acetate and sodium acetate in acetic acid. Solubility problems limited the accuracy of their results which, however, were consistent with a rate expression,

$$\text{Rate} = \frac{kK[Na_2Pd(OAc)_4][C_2H_4]}{[NaOAc]^2}$$

where K, the equilibrium constant for the reaction,

$$C_2H_4 + Na_2Pd(OAc)_4 \rightleftharpoons C_2H_4Pd(OAc)_2 + 2NaOAc$$

was estimated to be 3.1 ± 1 mole/liter. However, it is not yet certain that palladium acetate does exist under these conditions as $Na_2Pd(OAc)_4$.

One further problem which arises in the work which has been carried out in chloride-free media is that oxygen (even in trace amounts) can influence the course of these reactions very profoundly. This has been demonstrated by Davidson and co-workers,[225] but earlier workers probably did not allow for this.

van Helden *et al.*[285] (using palladium acetate containing some nitrogenous impurities) found that the rate of ethylene consumption was first-order in palladium acetate and in sodium acetate and suggested that the reactive species was $Pd(OAc)_3^-$. The reaction was presumed to proceed by an S_N2-type displacement of hydride by coordinated acetate.

The rate of the overall reaction was strongly catalyzed by lithium chloride; under these conditions, however, the selectivity toward vinyl acetate formation decreased and acetaldehyde formation became more important. The amount of ethylidene diacetate remained the same. At $[Cl^-]:[Pd(II)]$ ratios greater than 2, however, acetoxylation was completely suppressed and dimerization of the ethylene to butenes was the main reaction. This was ascribed to the ease of replacement of coordinated acetate by chloride in the palladium complex. A mixture of ferric acetate and lithium chloride had a greater effect on the rates than either component alone, and showed improved selectivity toward vinyl acetate. At high ethylene pressures under these conditions, formation of ethylidene diacetate was suppressed, but dimerization to butenes did occur. Ethylidene diacetate was the major product (46%) at higher temperatures (>80°) and at low water concentrations using a catalyst of Pd(II), Fe(III), Li^+, Cl^-, OAc^-, and HNO_3 with oxygen and C_2H_4 at 1 atm and 90–95°. At much higher nitrate concentrations the glycol mono- and diacetates were the main products.[285, 297, 298]

Clark *et al.* have reported that the use of amides, especially DMAC in the presence of copper(II), led to a greater selectivity for vinyl acetate formation.[290, 361] These results are reproduced in Table II-1.

TABLE II-1

Effect of DMAC on the Oxidation of Ethylene in Acetic Acid[a, b]

Volume fraction of acetic acid in DMAC	Yield of products (%)		
	$CH_2{=}CHOAc$	CH_3CHO	$CH_3CH(OAc)_2$
0	95	2	0
0.25	90	7	1
0.50	80	14	2
0.75	32	55	8
0.95	16	42	34

[a] After Clark *et al.*[361]
[b] Catalyst initially: $PdCl_2$ (0.1 *M*), LiCl (0.2 *M*), LiOAc (0.4 *M*), $Cu(OAc)_2$ (0.15 *M*); $[C_2H_4]/[O_2] = 2$; 1 atm, 100°).

Schultz and Rony[288] have suggested that the major effect of DMAC was to repress the decomposition of vinyl acetate with acetic acid (to acetaldehyde and acetic anhydride). DMSO, esters, trialkylamines, and particularly nitriles were also very effective; it was thought that each of these complexed more strongly with Pd(II) than vinyl acetate but less strongly than ethylene. Hence the formation of vinyl acetate was less inhibited than its decomposition (Section C,4).

The effect of a number of variables on the relative amounts of acetaldehyde, vinyl acetate, and ethylidene diacetate has been reported by Tamura and Yasui.[293] They estimated the activation energy for the formation of vinyl acetate at 16.9 kcal/mole, and suggested that the ethylidene diacetate also arose directly from ethylene.

The simplest mechanism for the formation of vinyl acetate, based on analogies to other systems, is the conversion of an ethylene π complex, by insertion of ethylene into a Pd–OAc bond (or by attack by external OAc^- on the coordinated olefin), into a σ complex (**II-6**). This can then decompose by β-elimination of H-PdOAc (however, see section B,5).

$$C_2H_4 + LPd(OAc)_3^- \rightleftharpoons \left[(CH_2{=}CH_2)(AcO)Pd(OAc)_2 \right]^- \overset{(L)}{\rightleftharpoons}$$

$$\left[AcOCH(H){-}CH_2{-}Pd(OAc)(AcO)(L) \right]^- \longrightarrow AcOCH{=}CH_2 + [HPd(OAc)_2(L)]^-$$

(**II-6**)

$$[HPd(OAc)_2(L)]^- \rightarrow HOAc + Pd^0 + OAc^- + (L)$$

(L = solvent, etc.)

Evidence for this reaction scheme and for the intermediacy of the σ complex (**II-6**) comes from the work of Moiseev and Vargaftik.[324] They showed that β-acetoxyethylmercuric chloride reacted with palladium chloride in ether to give vinyl acetate and with palladium chloride in acetic acid, in the presence of acetate, to give ethylidene diacetate.

$$AcOCH_2CH_2HgCl + PdCl_2 \xrightarrow{Et_2O} CH_2{=}CHOAc$$

$$AcOCH_2CH_2HgCl + PdCl_2 \xrightarrow{HOAc/OAc^-} CH_2{=}CHOAc + MeCH(OAc)_2$$

The formation of ethylidene diacetate can be accommodated in the reaction scheme if it is supposed that the σ complex (**II-6**) is able to decompose by two routes. One, as shown, is by β elimination of H–Pd, the other alternative is for isomerization to the potentially more stable† α-acetoxyethyl complex **(II-7)** to occur.

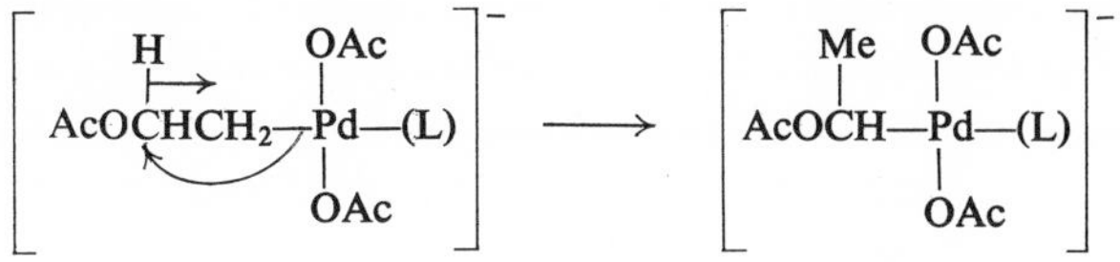

$$\left[\text{AcOCH(H)CH}_2\text{—Pd(OAc)}_2\text{—(L)}\right]^- \longrightarrow \left[\text{AcOCH(Me)—Pd(OAc)}_2\text{—(L)}\right]^-$$

(II-6) **(II-7)**

$$\left[\text{AcOCH(Me)—Pd(OAc)}_2\text{—(L)}\right]^- + \text{OAc}^- \longrightarrow (\text{AcO})_2\text{CHMe} + \text{Pd}^0 + 2\text{OAc}^- + (\text{L})$$

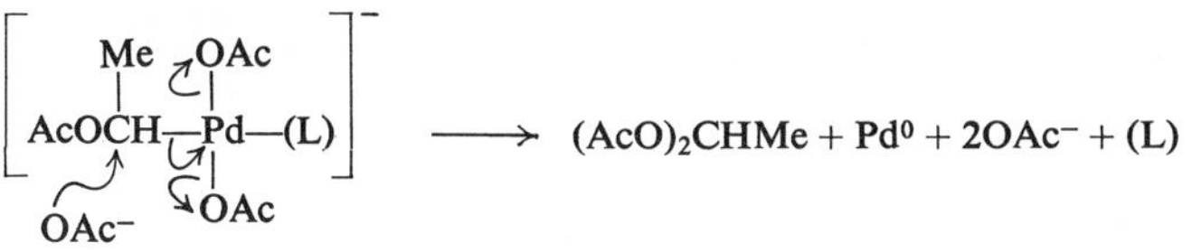

(L = solvent or ligand)

This could now react to give the product, either by attack of acetate at the α carbon, or by an intramolecular rearrangement to give the same products. Since acetate is potentially a bidentate ligand this appears the more likely.

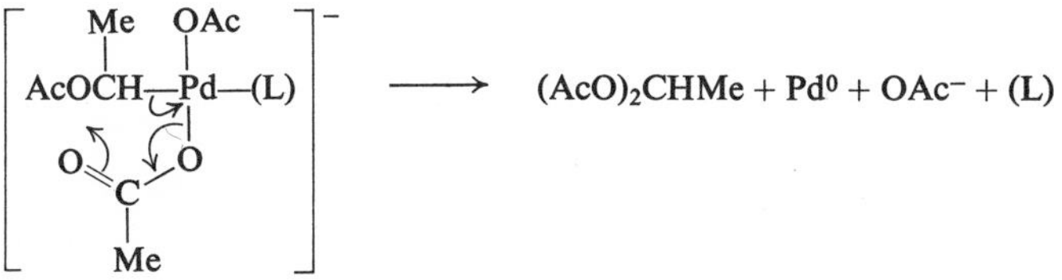

$$\left[\text{AcOCH(Me)—Pd(OAc)(O}_2\text{CMe)—(L)}\right]^- \longrightarrow (\text{AcO})_2\text{CHMe} + \text{Pd}^0 + \text{OAc}^- + (\text{L})$$

A further possibility, especially in the copper-catalyzed reactions at high $[Cl^-]:[Pd(II)]$ ratios, where the other coordination sites on the metal are occupied by chloride rather than acetate ligands (see below), is that the Pd–C bond is labilized toward heterolytic cleavage after isomerization to give a relatively stable intermediate cation, $[MeCO\cdot O{=}CHMe]^+$, which then undergoes nucleophilic attack.

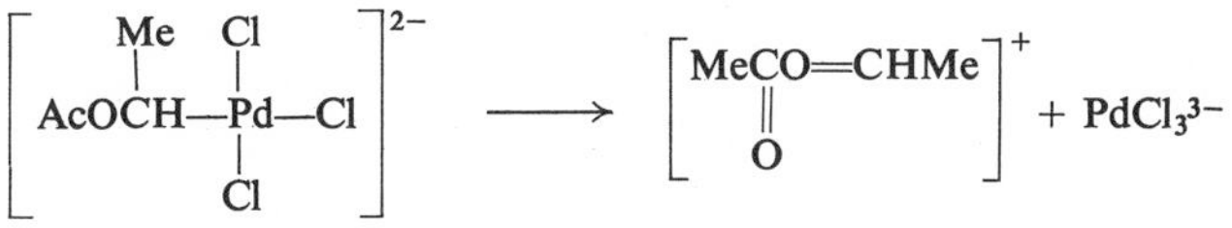

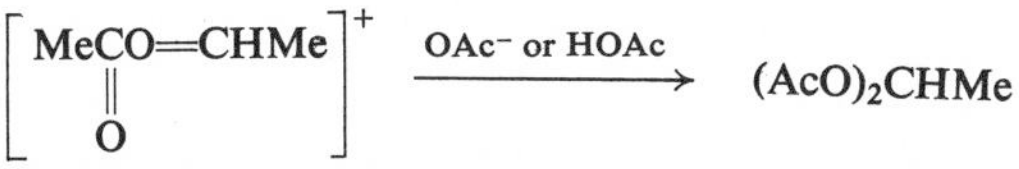

$$\left[\text{AcOCH(Me)—PdCl}_3\right]^{2-} \longrightarrow \left[\text{MeC(=O)O=CHMe}\right]^+ + \text{PdCl}_3{}^{3-}$$

$$\left[\text{MeC(=O)O=CHMe}\right]^+ \xrightarrow{\text{OAc}^- \text{ or HOAc}} (\text{AcO})_2\text{CHMe}$$

† Since the carbon to which the metal is attached now also bears an electron-withdrawing group.

Furthermore, since chloride is known to assist isomerization, it probably also accelerates the reaction to give the isomerized intermediate.

In practice these reactions were usually run in the presence of copper(II) salts to reoxidize the palladium. Under these conditions, however, new reactions arose which were very probably due, in this case, to cocatalysis by copper. In particular, $AcOCH_2CH_2OAc$, $AcOCH_2CH_2OH$, and $AcOCH_2CH_2Cl$ were obtained, which were not formed under *these* conditions in the absence of Cu(II).

However, Tamura and Yasui[295,297–299] have shown that the glycol mono- and diacetates *were* formed in the absence of copper if a large excess of lithium nitrate was present. This reaction

$$C_2H_4 + HOAc + PdCl_2 + LiNO_3 \rightarrow AcOCH_2CH_2OAc + HOCH_2CH_2OAc + MeCHO$$

was catalytic and could be run in the presence or absence of oxygen. Lithium nitrate was the most effective cocatalyst [more so than Fe(III) or Cu(II) nitrates], no metal was precipitated during the reaction as the nitrate acted as reoxidant, being itself reduced to nitric oxide.[297] van Helden *et al.*[285] have briefly reported a similar reaction. The rate in the absence of oxygen was first-order in both ethylene and palladium and an activation energy of 11.8 kcal/mole was calculated.[298] Under optimum conditions a 95% yield of the mono-acetate ($HOCH_2CH_2OAc$) was obtained,[297] and it appeared that the diacetate ($AcOCH_2CH_2OAc$) was formed by esterification of the monoacetate.[302]

The copper-catalyzed reactions in the absence of nitrate have been studied by Henry,[300] Clark *et al.*,[361–363] and Tamura and Yasui.[302] A serious problem in these studies has been the lack of solubility of the inorganic salts in the reaction medium, acetic acid.

Henry found that an acetic acid solution of LiCl (2 *M*), LiOAc (0.5 *M*), and Pd(II) (0.1 *M*) slowly oxidized ethylene to vinyl acetate. On addition of 0.1 *M* of copper(II) chloride, the rate of oxidation was increased considerably and the products were now the diacetate and the chloroacetate ($ClCH_2CH_2OAc$) almost exclusively. The copper was reduced to the (I) state and palladium was not precipitated, in contrast to the reaction in the absence of copper. Henry also showed that vinyl acetate was not an intermediate in the formation of $AcOCH_2CH_2OAc$ and $ClCH_2CH_2OAc$, and that the ratio of these two compounds to vinyl acetate increased with increasing Cu(II) concentration, whereas the yield of vinyl acetate increased at higher temperatures and pressures.

In the presence of water, the monoacetate ($HOCH_2CH_2OAc$) was formed; under optimum conditions the yield was over 90%.[302]

Clark *et al.*[361,363] obtained similar results to those of Henry for the reaction of ethylene and oxygen with a palladium acetate (0.005 *M*), lithium acetate

(0.5 *M*), and copper acetate (0.15 *M*) catalyst in acetic acid. On addition of lithium chloride (the most soluble chloride) selectivity toward vinyl acetate decreased, first to give more ethylidene diacetate and acetaldehyde and later, at higher chloride ion concentrations, to give glycol esters and $ClCH_2CH_2OAc$ (Table II-2). The rate of the overall reaction was first-order in ethylene and copper(II), and approximately first-order in palladium(II). A complex dependence on lithium chloride concentration was observed, approximately to the power -1 in uncomplexed chloride.

Obviously considerable work is still necessary to define precisely the mechanisms of these reactions, and, in particular, the precise function of the copper. There are two possible explanations for the change in product composition at high $[Cl^-]$ concentration in the presence of copper. One is that a trans-alkylation occurs to give an alkylcopper complex, and the other is that a mixed palladium–copper chloride complex is formed in which the Pd–C bond is retained, but is now activated toward nucleophilic attack.

Clark *et al.*[361, 363] have shown that at high chloride ion concentrations the palladium is largely complexed by chloride rather than acetate. If a mixed palladium–copper complex, for example, (**II-8**), is formed and the Pd–C bond is more reactive, two reaction paths are possible.

(a) The complex (**II-8**) can decompose to $AcOCH_2CH_2X$ ($X = Cl$, OH, AcO), by direct attack at the α carbon by X^-.

(b) Alternatively, as suggested by Clark *et al.*,[361] the Pd–C σ bond undergoes heterolytic cleavage, assisted by nucleophilic attack (by X^-) at the metal, to give $AcOCH_2CH_2^+$, which is stabilized as the cyclic acetoxonium ion (**II-9**). This intermediate can then react with whichever nucleophile is present in greatest abundance (or with AcOH) allowing for the relative nucleophilicities of the various species.

$$\left[AcOCH_2CH_2{-}Pd(\mu\text{-}Cl)_2Cu(Cl)_2\right]^- \xrightarrow[(a)]{X^-} AcOCH_2CH_2X + [PdCuCl_4]^{2-}$$

(II-8)

(b) ↓ L

$$[LPdCuCl_4]^{2-} + \text{(cyclic } CH_2{-}CH_2 \text{ acetoxonium ion, } O{-}\overset{+}{C}H(Me){-}O) \xrightarrow{X^-} AcOCH_2CH_2X$$

$$\xrightarrow{AcOH} AcOCH_2CH_2OAc + H^+$$

(II-9)

(L = solvent, Cl^-; X = OH, Cl, OAc)

TABLE II-2

Effect of Chloride Concentration on the Products Formed in the Catalytic Oxidation of Ethylene in Acetic Acid[a,b]

Molarity of chloride	Yield (%)					
	$CH_2{=}CHOAc$	MeCHO	$MeCH(OAc)_2$	$AcOCH_2CH_2OAc$	$ClCH_2CH_2OAc$	$HOCH_2CH_2OAc$
0.0	80	4	1	0	0	0
0.025	55	30	2	0	0	0
0.055	43	34	3	0	0	0
0.075	41	35	4	0	0	0
0.125	40	40	5	0	0	0
0.225	34	25	25	1	0	0
0.525	11	23	32	18	2	1
1.025	5	15	16	42	6	2
1.525	1	5	4	56	11	3
2.025	1	2	2	55	14	5

[a] After Clark *et al.*[361]

[b] Catalyst initially: $Pd(OAc)_2$ (0.005 *M*), LiOAc (0.5 *M*), $Cu(OAc)_2$ (0.15 *M*), at 105° and 1 atm; $[C_2H_4]:[O_2] = 2$.

The observation that similar products are formed in the presence of nitrate, without copper, suggests that in the presence of suitable oxidizing ligands the Pd–C σ bond can always be activated toward heterolytic cleavage, and that a transalkylation does not occur. A mechanism has been suggested for the nitrate-promoted reaction by Tamura and Yasui[298]; the activation energy for this reaction was found to vary with the [LiCl]: [Pd(OAc)$_2$] ratio and was 14 kcal/mole for a ratio of 6 and 24 kcal/mole in the absence of LiCl.

Shimizu and Ohta[364] observed an induction period in the copper-cocatalyzed vinyl acetate synthesis when $[Cl^-] > [OAc^-]$. During the induction period $ClCH_2CH_2OAc$, $HOCH_2CH_2OAc$, and $AcOCH_2CH_2OAc$ were formed; at steady state the products were vinyl acetate, acetaldehyde, and ethylidene diacetate.

In the vinyl acetate synthesis, copper is present as a reoxidant and chloride is necessary to solubilize various intermediates in the reoxidation. The problems associated with the reoxidation stage in this reaction have been discussed by Bryant and MeKeon.[365]

Patents on the homogeneous vinyl acetate synthesis include References 101, 280, 281, 291, 292, 307, 366–386.

Nakamura and Yasui[273] have discussed the Kurashiki Rayon Co. heterogenerously catalyzed synthesis of vinyl acetate. The catalyst was, effectively, finely divided palladium on an inert carrier, together with a promoter (KOAc was found to be the best). Under the conditions of the reaction (C_2H_4 + HOAc + O_2 at 120°) palladium acetate was formed. However, the mechanism suggested involved the formation of vinylpalladium at the surface which was then attacked by coordinated acetate. Both this reaction and that with propylene (to give only allyl acetate) were very specific.

5. Oxidation of Higher Olefins in Acetic Acid

The situation here is extremely complex and, at the time of writing, quite unresolved. For example, the acetoxylation of cyclohexene has been studied by a number of workers. Green *et al.*[387] and Henry[388] obtained the cyclohex-2-enyl acetate **(II-11)** (76%) and the cyclohex-3-enyl acetate **(II-12)** (24%)

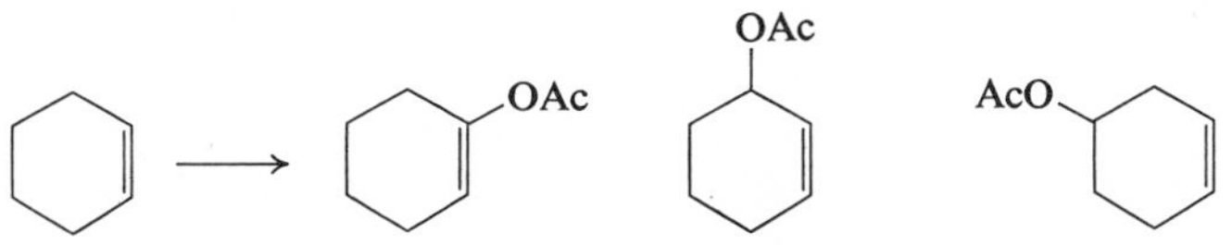

OAc OAc AcO

(II-10) **(II-11)** **(II-12)**

using palladium chloride and acetate in acetic acid. Anderson and Winstein[389] using palladium acetate in acetic acid, obtained only the allylic acetate (**II-11**). All these workers agreed that none of the enol acetate (**II-10**) was formed. However, Brown *et al.*[225] later showed that in the absence of oxygen (**II-10**) was the major product of the palladium acetate reaction in the presence of perchloric acid. In the absence of oxygen and under neutral or basic conditions (no chloride present) these workers found that no acetates were formed; benzene, arising by dehydrogenation of the cyclohexene was the main product. They also found that palladium acetate reacted with cyclohexene in the presence of oxygen to give (**II-11**) and (**II-12**) as well as cyclohex-1-en-3-ol and cyclohex-1-en-3-one; (**II-11**) was the major product in the presence of oxygen and quinone.[390] (See also Appendix.)

OAc

AcO

Pd(II)/O_2/HOAc

+

(II-11) **(II-12)**

OAc

Pd(II)/H^+

(II-10)

This suggests the need for considerable caution in interpreting some of these results. The effect of oxygen does not appear to be as important in the presence of chloride ion; a side effect of chloride, however, is to increase the degree of isomerization (H transfer) that occurs.

Propylene reacted with acetate in the presence of palladium(II) to give *cis*- and *trans*-1-acetoxypropene, 2-acetoxypropene, and 3-acetoxypropene.[62, 304, 361, 372, 391–398] Again these were obtained in differing amounts, depending on conditions; in a chloride-free medium, Kitching *et al.*[397] observed 98.6% formation of 2-acetoxypropene (isopropenyl acetate). Stern obtained 64% of this isomer and 36% of the 1-acetoxypropene,[393] while Belov *et al.*, under slightly different conditions, but again using palladium chloride,[392] obtained 44% 2-, 36% *cis*-1, 17% *trans*-1-, and 3% 3-acetoxy-1-propene. However, an ICI patent claimed a 75% yield of the 3-isomer (allyl acetate) in a chloride-free medium in the presence of oxygen.[396] Other products which arose from this reaction included acrolein,[394, 395, 398] acetone,[394, 398, 399] isopropyl acetate,[372] propylidene diacetate,[392] and, particularly in the presence of copper(II), 1,3-diacetoxypropane.†[400] Tamura and Yasui[295, 298] have

† 1-Methyl-3-ethylallylpalladium chloride dimer has also been reported as a by-product.[388]

obtained 1,2-glycol monoesters using palladium chloride in the presence of nitrate in acetic acid. A gas-phase acetoxylation has also been reported.[399] (See also Nakamura and Yasui.[273])

Stern showed that under his conditions ($PdCl_2/HPO_4^{2-}/HOAc$/isooctane) propylene reacted 2.8 times as fast as 2-deuteropropylene, and that the acetates derived from the latter (1- and 2-acetoxy-1-propenes) retained 75% of their deuterium.[393] This suggested that rupture of the C–H (D) bond was rate-determining and that a 1,2-shift of hydride from the carbon bearing the acetate group occurred, followed by loss of a proton from the adjacent carbon. Confirmation of this result is needed since, if correct, it would necessitate modification of the reaction scheme suggested above for the formation of vinyl acetate.

Clark *et al.*[361] noted that in a chloride-free medium the amount of allyl acetate increased with increase in the acetate concentration and that in the presence of chloride (and oxygen) the relative amounts of 1-, 2-, and 3-acetoxy-1-propenes, acrolein, and acetone varied considerably on changing solvent from acetic acid to DMAC.

Very careful studies on the effects of all variables on the product composition are necessary before more mechanistic speculation is embarked upon.

This sentiment is also true of work on the higher olefins. 1-Butene has been examined by Belov and Moiseev,[401] Kitching *et al.*,[397] and Henry[300] and is the subject of a patent.[400] In the absence of chloride, Kitching *et al.*[397] again found largely (80%) the 2-enol acetate, whereas in the presence of chloride five or six products were observed by Belov and Moiseev.[401] Henry showed that in, the presence of copper(II), nine glycol diacetates and chloroacetates were formed.[300] He suggested that these arose directly from a primary oxypalladation product by a copper-assisted heterolysis, as discussed above.

2-Butenes gave rise to only slightly less complex mixtures,[300, 400] although Kitching *et al.* claimed that in a chloride-free medium, 3-acetoxy-1-butene was the major product.[397] These latter workers also reported that under their conditions 1-olefins gave largely enol acetates, and 2-olefins largely allylic acetates. They rationalized their results in terms of an oxypalladation followed by loss of H–Pd. Belov and Moiseev again obtained large amounts of allylic acetates in the presence of chloride.[401]

Schultz and Gross have studied the reaction of 1-hexene with acetate under various conditions.[402] They observed that attack by acetate at the 2-position was favored at lower, and attack at the 1-position at higher, chloride ion levels.† "High boilers" (largely 1,2-glycol esters) were only formed at high chloride-to-acetate ratios in the presence of copper(II). Levanda and Moiseev[403] found that the acetoxylation of 1-hexene by $PdCl_2$ in the absence of copper gave 5-acetoxy-2-hexene, 2-acetoxy-3-hexene, and 4-

† The positions of double bonds were not determined.

acetoxy-2-hexene in the ratio 7:2:1. Chloro(1-ethyl-3-methyl-π-allyl)palladium dimer was obtained as a by-product.

Long-chain α-olefins have also been reported to give 1-acetoxy-2-alkenes[62, 290, 404, 405] and 1-acetoxy-1-alkenes[290]; butadiene gave butadienylacetate[304]; and allyl acetate gave glyceryl di- and triacetates.[406] Matsuda *et al.* have reported the acetoxylation of octenes, 1-methyl-cyclopentene, and 1-methylcyclohexene,[405] as well as of a variety of functionally substituted olefins, ethyl acrylate, acrylonitrile, allyl cyanide, etc.[407]

Baird showed that norbornene gave 50–84% *exo*-2-chloro-*syn*-7-acetoxynorbornane, **(II-13)**, and has rationalized this in terms of acetoxypalladation followed by heterolysis of the C–Pd bond to give a cation which

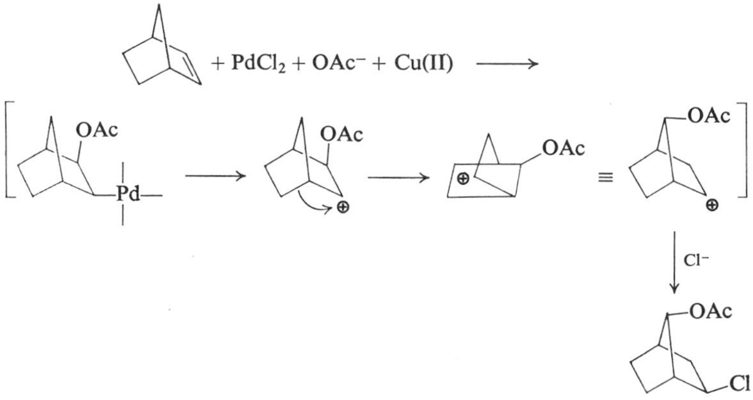
+ $PdCl_2$ + OAc^- + Cu(II)
OAc
Pd
OAc
OAc
OAc
Cl⁻
OAc
Cl

(II-13)

rearranged.[408] Analogous reactions have been reported by Battiste[409] and Baker[410] and their co-workers. The latter workers interpreted their results as not involving the formation of a free carbonium ion during the reaction. Nucleophilic attack on the Pd–C σ bond was postulated. These reactions did not occur in the absence of Cu(II) and are again examples of the activation of the Pd–C bond to nucleophilic attack in the presence of (other) oxidizing ligands.

Styrene also underwent acetoxylation. Moritani and his co-workers[232,233,235] isolated α-acetoxyethylbenzene, **(II-14)**, and *cis*-β-acetoxystyrene, **(II-15)**, as by-products in the reaction of the styrenepalladium chloride complex with aromatic hydrocarbons in the presence of acetic acid.

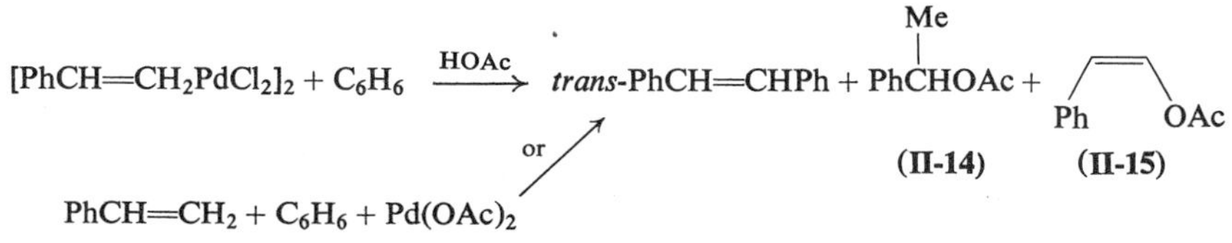
[PhCH=CH₂PdCl₂]₂ + C₆H₆
HOAc
trans-PhCH=CHPh + PhCHOAc +
Me
Ph
OAc
or
PhCH=CH₂ + C₆H₆ + Pd(OAc)₂

(II-14) **(II-15)**

In the presence of copper(II) Uemura and Ichikawa[411] isolated some ten products including the above, acetophenone, and chloro-, dichloro-, chloro-acetoxy-, diacetoxy-, and hydroxyacetoxyethylbenzenes. These workers also obtained **(II-16)** and **(II-17)** from styrene and acetylacetone, catalyzed by palladium(II) and copper(II).[412]

$$PhCH{=}CH_2 + MeCOCH_2COMe \xrightarrow{Pd(II)/Cu(II)}$$

MeCO, Ph, Me, O — **(II-16)** + MeCO, Me, O, Ph — **(II-17)**

Heck[9] has shown that the products from reaction of "phenylpalladium acetate" and olefins can be acetoxylated with lead tetraacetate.

$$RCH{=}CH_2 + PhHgOAc + Pd(OAc)_2 + Pb(OAc)_4 \longrightarrow RCH(OAc)CH_2Ph$$

$(R = Me, CH_2{=}CH)$

Shier[7, 133] reported that allene reacted with palladium acetate (or nitrate) in acetic acid in the presence of sodium acetate to give allyl acetate. This was a minor product (11 %), however; the major products arose from dimerization of allene (see Volume I, Chapter V, Section F,12 and this Volume, Chapter I, Section C,3,b).

The diacetoxylation of butadiene (giving 1,4-diacetoxy-2-butene and 3,4-diacetoxy-1-butene) and piperylene has been reported in an ICI patent.[413] (See also Stern and Spector.[304]) 1,3-Cyclohexadiene gave largely 1,4-di-acetoxycyclohex-2-ene and 1,4-cyclohexadiene gave 1,3-diacetoxycyclohex-4-ene, with palladium acetate in the presence of oxygen.[390] Benzene was the major product in the absence of oxygen.

At present, therefore, while mechanisms to explain the various products arising from higher olefins and acetate in the presence of palladium(II) can and have been written, a much more careful study of these reactions than has hitherto been carried out is necessary before such suggestions can be critically evaluated.

6. Oxidation of Olefins in Alcohols: The Formation of Vinyl Ethers, and Acetals

Moiseev *et al.*[303] and Stern and Spector[304] in 1961 independently reported the reaction of the ethylenepalladium chloride complex with alcohols in the presence of base to give vinyl ethers and acetals.† The latter were preferentially formed at lower temperatures.

$$[C_2H_4PdCl_2]_2 + ROH + base \rightarrow CH_2{=}CHOR + CH_3CH(OR)_2 + Pd + base \cdot HCl$$

Isolation of the olefin complex is not necessary and the reactions are the subject of a number of patents since they can be run catalytically in the presence of a copper(II) salt and oxygen.[101, 102, 276–278, 378, 414]

By-products included butenes,[101, 102] acetaldehyde,[101, 102, 278] esters,[102, 414] and β-chloroethyl ether (especially at high copper chloride concentrations).[276]

Moiseev and Vargaftik[294] showed that the acetal, $MeCH(OMe)_2$, prepared in CH_3OD contained virtually no deuterium. Therefore this product could not have arisen from addition of methanol to vinyl methyl ether.

These authors later[324] showed that β-ethoxyethylmercuric chloride reacted with palladium chloride in ether to give ethyl vinyl ether.

$$EtOCH_2CH_2HgCl + PdCl_2 \xrightarrow{Et_2O} \left[EtOCH_2CH_2\overset{|}{\underset{|}{Pd}}— \right] \longrightarrow$$

(II-18)

$$EtOCH{=}CH_2 + Pd + HgCl_2 + HCl$$

This suggests again that an intermediate such as **(II-18)** arising from ethoxy-palladation of ethylene may be present in the reactions leading to vinyl ethers and acetals.

A detailed study of these reactions has been reported by Ketley and Fisher.[306] They reacted $[C_2H_4PdCl_2]_2$ with a number of alcohols at 20° in the presence of Na_2HPO_4 under conditions such that the acetals were the only significant products. They showed that, of the alcohols they tried, reaction in the above sense occurred only for methanol and ethanol. Ethylene was liberated in all reactions, and particularly with the higher alcohols. No acetals derived from propylene or butene were obtained on reaction with complexes of these olefins. They explained this by assuming that the alcohols under these conditions merely replaced the olefin to give $(ROH)PdCl_2$ complexes which then decomposed to give aldehydes, acetals, and alkyl chlorides derived from R (Section E). This reaction even occurred in the reactions of $[C_2H_4PdCl_2]_2$ with methanol and ethanol. Using ^{14}C-labeled ethylene in the reaction of $[C_2H_4PdCl_2]_2$ with ethanol they showed that some 15% of the diethyl acetal obtained was derived from ethanol, not ethylene.

Once again, therefore, a reaction pathway involving oxypalladation, followed by (a) β elimination of H–Pd, or (b) isomerization followed by heterolysis can be envisaged here.

† Ketley and Braatz[305] noted that in the *absence* of base (HPO_4^{2-}), $[C_2H_4PdCl_2]_2$ reacted with isopropanol to give isopropylchloride and ethylene only.

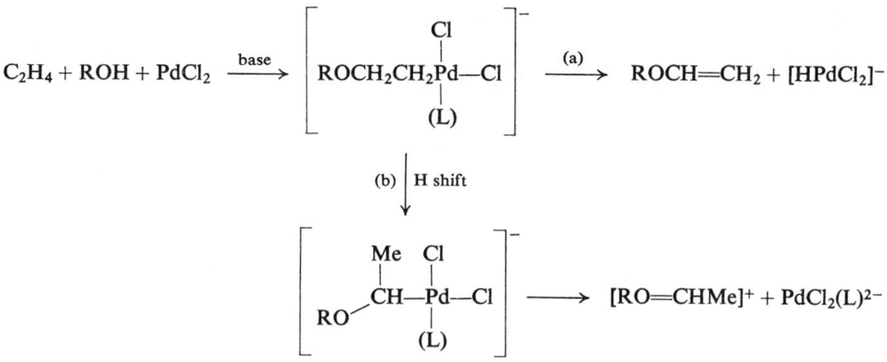

$$C_2H_4 + ROH + PdCl_2 \xrightarrow{\text{base}} \left[ROCH_2CH_2Pd(Cl)(L){-}Cl \right]^- \xrightarrow{(a)} ROCH{=}CH_2 + [HPdCl_2]^-$$

(b) ↓ H shift

$$\left[RO{-}CH(Me){-}Pd(Cl)(L){-}Cl \right]^- \longrightarrow [RO{=}CHMe]^+ + PdCl_2(L)^{2-}$$

$$[HPdCl_2(L)]^- \rightarrow HCl + Cl^- + Pd + (L)$$

$$[RO{=}CHMe]^+ + ROH \rightarrow (RO)_2CHMe + H^+$$

$$PdCl_2(L)^{2-} \rightarrow Pd + 2Cl^- + (L)$$

(L = solvent, etc.)

Further investigations will no doubt show how closely related these reactions are to the vinyl acetate–ethylidene diacetate reactions. Similar effects of solvents and halide ions on the products may be expected.

François[332] has studied the kinetics of the copper-cocatalyzed reaction of ethylene and methanol. As for the reaction in water, he has interpreted his results here in terms of a reaction {at high [Cu(II)]} involving copper and one predominating at low [Cu(II)] not involving copper. The rate of the former was given by

$$\frac{k[PdCl_4^{2-}][C_2H_4][CuCl_3^-]^2}{[H^+][Cl^-]}$$

No deuterium was incorporated into methyl vinyl ether prepared in MeOD.

Lloyd and Luberoff[415] have studied the reactions of higher olefins with alcohols or ethylene glycol catalyzed by palladium and copper chlorides. Ethylene glycol always gave the 1,3-dioxolane (ketal), propanol always the ketone, and ethanol and methanol, under their conditions, usually but not invariably, the ketone.

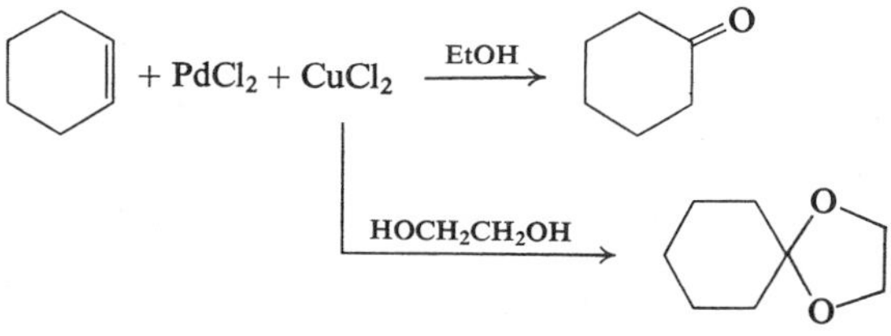

Cyclohexene + $PdCl_2$ + $CuCl_2$ $\xrightarrow{\text{EtOH}}$ cyclohexanone

Cyclohexene + $PdCl_2$ + $CuCl_2$ $\xrightarrow{HOCH_2CH_2OH}$ cyclohexanone ethylene ketal (1,3-dioxolane)

These workers also noted that the reactions proceeded more readily in alcohols than in water, and that the latter, in fact, inhibited the reaction. For example, an 88% conversion of cyclohexene (giving 58% cyclohexanone) was observed in ethanol without water; addition of 10% water by volume reduced the conversion to 48% and gave only 38% cyclohexanone.

C. THE PALLADIUM-CATALYZED SUBSTITUTION OF Y FOR X IN $XCH{=}CH_2$

A large number of reactions, strictly catalytic in Pd(II) (no reduction to metal occurs), are known in which nucleophilic replacement of a vinylic substituent (other than H) can occur. They are of the general form:

$$CH_2{=}CHX + Y^- \xrightarrow{Pd(II)} CH_2{=}CHY + X^-$$

These reactions were first reported by Smidt *et al.*[416, 417] where X and Y were carboxylates.

$$RCOOH + R'COOCH{=}CH_2 \xrightarrow{Pd(II)} RCOOCH{=}CH_2 + R'COOH$$

Competition with a decomposition reaction also occurred.†

$$CH_2{=}CROCOR' + HOR'' \xrightarrow{Pd(II)} CH_3COR + R'COOR''$$
$$(R = H, Me \text{ and } R'' = H, Ac, \text{etc.})$$

Vinyl chloride underwent rather similar reactions.[414, 418–422]

$$CH_2{=}CHCl + OAc^- \rightarrow CH_2{=}CHOAc + Cl^-$$

Catalysts which have been used for this reaction include palladium chloride and HPO_4^{2-} or acetate, and palladium acetate in DMF. Alcohols also reacted.[383, 420, 421, 423]

$$CH_2{=}CHOAc + ROH \xrightarrow{Pd(II)} CH_2{=}CHOR + MeCHO + MeCH(OR)_2$$
$$(R = Me, Bu^t)$$

$$CH_2{=}CHCl + ROH \xrightarrow{Pd(II)} CH_3CH(OR)_2$$

† The palladium(II)-catalyzed hydrolysis of vinyl chloride to acetaldehyde was also reported by Smidt *et al.*[416]

Exchange reactions such as the following have also been reported[420, 424] (Chapter IV, Section A,5 and B).

$$CH_2{=}CHCl + F^- \xrightarrow{Pd(II)} CH_2{=}CHF$$

$$CH_2{=}CHCl + BuNH_2 \xrightarrow{Pd(II)} CH_2{=}CHNHBu$$

1. The Ester Exchange (Transvinylation) Reaction

Both Henry[425] and Smidt and co-workers[426, 427] have studied the mechanism of the ester-exchange reaction. Their results and mechanistic proposals agree. Henry, working in acetic acid with a palladium chloride–lithium chloride–lithium acetate catalyst found that the exchange

$$CH_2{=}CHOCOCD_3 + HOCOCH_3 \rightarrow CH_2{=}CHOCOCH_3 + HOCOCD_3$$

was first-order in vinyl acetate and lithium acetate. Lithium chloride inhibited the reaction strongly and he suggested a dependence on $[Li_2Pd_2Cl_6]^1$. Using *cis*- and *trans*-propenyl acetates he was able to show that isomerization was always accompanied by incorporation of $OCOCD_3$ when it was carried out in CD_3COOH.

$$(H)(OCOCH_3)C{=}C(H)(CH_3) + CD_3COOH \longrightarrow (H)(OCOCD_3)C{=}C(CH_3)(H) + CH_3COOH$$

Jira *et al.*[426, 427] showed that the *cis*- and *trans*-propenyl acetates reacted with propionate to give the propenyl propionates. The exchange reaction was always accompanied by isomerization.

$$(H)(OCOMe)C{=}C(H)(Me) + EtCOO^- \longrightarrow (H)(OCOEt)C{=}C(Me)(H) + MeCOO^-$$

$$(H)(OCOMe)C{=}C(Me)(H) + EtCOO^- \longrightarrow (H)(OCOEt)C{=}C(H)(Me) + MeCOO^-$$

This reaction was followed by a slower isomerization of the propenyl propionates.

Using ^{18}O-labeled acetic acid, Sabel *et al.*[427] showed that the labeled oxygen was entirely incorporated into vinyl acetate in the exchange reaction

$$CH_2{=}CHOCOEt + MeC^{18}O_2H \rightarrow CH_2{=}CH^{18}OC^{18}OMe + EtCOOH$$

Therefore the $CH_2{:}CH{-}O$ bond was the one cleaved. They termed these reactions *transvinylations*; in contrast, organic ester hydrolysis and transesterification reactions usually proceed by acyl–oxygen cleavage (RCO⌇OR′).

The mechanism of the transvinylation reaction appears very similar to the phenylation of olefins described by Heck[25] (Chapter I, Section A). In this case too, the results are best explained by a cis oxypalladation of the unsaturated ester followed by a cis deoxypalladation.

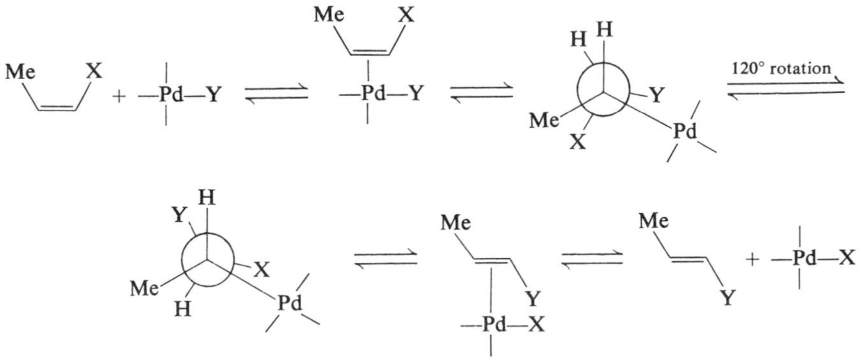

(X = OCOR; Y = OCOR′)

By the principle of microscopic reversibility, both addition and elimination must occur in the same sense. That both reactions occur trans is also conceivable. Either hypothesis explains the observation that one exchange is always accompanied by one isomerization and vice versa.

Henry also noted that 1-acetoxycyclopentene did not undergo ester exchange as would be expected on this mechanism since rotation of the oxypalladation intermediate cannot occur here.[425] (See also p. 112 and Appendix.)

Sabel *et al.* found that other metals [Hg(II), Pt(II), and Rh(III)] also catalyzed the ester exchange.[427]. The mercury-catalyzed reaction was also a transvinylation, but was cocatalyzed by acid, rather than base.

The closely related reaction,

$$CH_2{=}CHOBu + EtOH \rightleftharpoons CH_2{=}CHOEt + BuOH$$

occurs at low temperatures in the presence of $PdCl_2$. McKeon *et al.*[428] also showed that both *cis*- and *trans*-propenyl ethyl ethers reacted stereospecifically with propanol, with inversion. This again implies the occurrence of an oxypalladation-deoxypalladation mechanism. Above $-25°$ the $PdCl_2$ catalyst began to decompose to give HCl which then catalysed the formation of acetals,

$$CH_2{=}CHOBu + EtOH \rightarrow MeCH(OEt)_2 + MeCH(OBu)_2 + MeCH(OEt)OBu$$

This reaction could be eliminated by using stabilized palladium acetate complexes, for example (bipy)$Pd(OAc)_2$, as catalysts. Under these conditions reactions such as,

$$CH_2{=}CHOAc + EtOH \rightarrow CH_2{=}CHOEt + AcOH$$

were run at 100°. The same mechanism occurs here too.

2. The Esterification of Vinyl Chloride

Kohll and van Helden[421] found that normally very unreactive vinylic chlorines were easily exchanged by acetate in a very clean palladium-catalyzed reaction. Below 80° little metal was precipitated and the palladium chloride could be recovered. They found the reaction to be first-order in acetate ion and palladium(II), while chloride inhibited the reaction. The use of DMF as a solvent caused a twofold increase in rate; this was ascribed to the greater nucleophilic character of acetate in this solvent. However, no reaction occurred in DMSO, owing to the deactivation of the palladium chloride by formation of $(DMSO)_2PdCl_2$.

Volger[422] studied this reaction further using labeled vinyl chloride. 2,2-Dideuterovinyl chloride underwent the reaction without appreciable loss or scrambling of deuterium, hence no hydrogen transfer occurred.

$$CD_2{=}CHCl + OAc^- \xrightarrow{PdCl_2/HOAc} CD_2{=}CHOAc + Cl^-$$

Trans-2-deuterovinyl chloride underwent the reaction, without appreciable loss of deuterium, to give equal amounts of *cis*- and *trans*-2-deuterovinyl acetates. However, it was shown that isomerization of *trans*- to *cis*-2-deuterovinyl chloride occurred during the reaction,

$$\underset{\text{(D,H)C=C(H,Cl)}}{} \quad (D)(H)C{=}C(H)(Cl) \longrightarrow (D)(H)C{=}C(H)(OAc) + (H)(D)C{=}C(H)(OAc)$$

$$\downarrow \qquad \nearrow$$

$$(H)(D)C{=}C(H)(Cl)$$

Jira *et al.*[426] have studied the reactions of *cis*- and *trans*-propenyl chlorides with acetate in the presence of $PdCl_2$. They found that these reactions occurred stereospecifically with *retention* of configuration:

$$\begin{matrix} H & & Cl \\ & C & \\ & \| & \\ & C & \\ H & & Me \end{matrix} \xrightarrow{OAc^-/Pd(II)} \begin{matrix} H & & OAc \\ & C & \\ & \| & \\ & C & \\ H & & Me \end{matrix} \qquad \begin{matrix} H & & Cl \\ & C & \\ & \| & \\ & C & \\ Me & & H \end{matrix} \xrightarrow{OAc^-/Pd(II)} \begin{matrix} H & & OAc \\ & C & \\ & \| & \\ & C & \\ Me & & H \end{matrix}$$

It therefore appears that this reaction and that described by Volger proceed analogously, and that isomerization of the *trans*-2-deuterovinyl chloride to the cis isomer is one reason for the isolation of *cis*-2-deuterovinyl acetate in the previous example. Isomerization of the trans-2-deuterovinyl acetate to the cis isomer must also occur in the reaction.

Again the similarity to the other reactions makes an oxypalladation intermediate attractive. However, since the configuration is retained in these reactions, in contrast to the ester exchange, a cis-insertion–cis-elimination mechanism will not suffice here and the direction of insertion must be opposed to that of elimination. Since the conditions for this reaction are closely similar to that for the ester exchange, a cis insertion may well be the first step. Elimination from the intermediate must then be trans, for example,

Me Cl (cis-alkene) + —Pd—OAc ⟶ Me Cl (alkene coordinated to —Pd—OAc) ⟶ (Newman projection: Me, Cl, H, H, OAc, Pd) ⟶ Me OAc (cis-alkene) + —Pd—Cl

The reason for the outgoing group eliminating trans here is not clear. It may possibly be assisted by a further molecule of Pd(II) in a non rate-determining step and thus eliminate trans for steric reasons. Alternatively, it is possible that the initial acetoxypalladation is trans; evidence for attack by uncoordinated nucleophiles has been cited elsewhere (p. 78 and Appendix). In that case the elimination of Pd–Cl must be cis and this may arise from the high nucleophilicity of chloride for Pd(II). This point is not yet resolved and merits further investigation.

As discussed by Volger,[422] the isomerization of *trans*- to *cis*-2-deuterovinyl chloride may also result from a chloropalladation intermediate and thus be analogous to the transesterification reaction.

The esterification of vinyl chloride has also been studied by Tamura and Yasui.[429] Yamaji *et al.*[430] obtained *trans*-1,2-diacetoxyethylene from both *cis*- and *trans*-1,2-dichloroethylene.

Brady recently reported that allylic chlorides readily reacted with acetate in the presence of $PdCl_2$ in DMF.[431] A π-allylic complex intermediate could be

$$CH_2{=}CHCH_2Cl + OAc^- \xrightarrow{PdCl_2/DMF} CH_2{=}CHCH_2OAc + Cl^-$$

excluded and the author suggested an oxypalladation, followed by elimination of Pd–Cl, by analogy with the mechanisms discussed above. The conversion of allyl acetate to allyl methyl ether using $LiCl/PdCl_2$ in methanol has also been described, see p. 126.

3. Other Reactions of Vinyl Chloride

Kohll and van Helden[421] have shown that replacement of chloride in vinyl chloride is a general reaction. Addition of base was necessary for the reaction to proceed with alcohols. Under these conditions the acetal was the chief product (aldehyde, arising from oxidation of the alcohol, was also obtained).

$$CH_2{=}CHCl + ROH + PdCl_2 + \text{base} \rightarrow MeCH(OR)_2 + HCl + Pd$$

These authors proposed that the reaction proceeded via primary formation of the vinyl ether, but gave no evidence for this. The observation that the palladium-induced hydrolysis of vinyl chloride to acetaldehyde was *not* catalytic[416] and that palladium was precipitated, suggest that this and the alcoholysis reactions are quite complex. Rhodium and ruthenium chlorides also catalyze the solvolysis of vinyl chloride.[432]

4. Other Reactions of Vinyl Acetate

Clement and Selwitz[286] observed that vinyl acetate reacted with palladium chloride in the presence of acetate in acetic acid to give acetaldehyde and acetic anhydride. This reaction appeared to be unique for palladium; ethylidene diacetate was not an intermediate nor was the acetic acid being dehydrated.

$$CH_2{=}CHOAc + HOAc \xrightarrow[Cl^-/OAc^-]{Pd(II)} CH_3CHO + Ac_2O$$

The reaction was strongly catalyzed by lithium or copper(II) acetates in the presence of chloride and was insignificant when palladium acetate alone was used.

Two patents from the Kurashiki Rayon Co.[433,434] have described the reverse of this reaction, to give largely vinyl acetate and some ethylidene diacetate, again catalyzed by palladium(II) in the presence of lithium, copper, acetate, and chloride ions.

$$CH_3CHO + Ac_2O \rightarrow CH_2{=}CHOAc + CH_3CH(OAc)_2$$

Since the reaction discovered by Clement and Selwitz is the cause of considerable loss of vinyl acetate during its synthesis by the liquid-phase process,† it has attracted some attention.

Schultz and Rony[287,288] found that the rate of the reaction was dependent on palladium(II) and acetate concentrations, and was inhibited by weakly coordinating solvents and high concentrations of chloride ion. Chloride ion in low concentration was essential and water promoted the reaction. The inhibition of this reaction by weakly coordinating solvents (DMF, DMSO, trialkylamines, esters, and particularly nitriles) was ascribed to these solvents complexing the metal in preference to vinyl acetate. Schultz and Rony also suggested that the improvement in the selectivity toward vinyl acetate in its synthesis from ethylene, conferred by the presence of weakly coordinating solvents, arose largely from inhibition of this decomposition reaction. Since ethylene complexed more strongly, the synthesis of vinyl acetate was not impeded.

These workers also found that when vinyl acetate was reacted in CH_3COOD, deuterium only appeared as CDH_2CHO in the product. The mechanism suggested by the authors is reproduced in Scheme II-2; it involves an oxypalladation step which occurs via a cyclic intermediate. The initial active species is presumed to be $[AcOCH{=}CH_2PdCl_3]^-$ or $[AcOCH{=}CH_2PdCl_3PdCl_2]^-$.

Henry has also studied the mechanism and has shown it to be unrelated to the ester exchange reaction[425]; this observation would appear to necessitate some modification of the mechanism proposed in Scheme II-2, since the two reactions cannot then involve the same steps.

Kohll and van Helden[218] observed that vinyl acetate underwent acetoxylation and coupling when reacted with palladium acetate (see Chapter I, Section D).

As already mentioned (Chapter I, Section A) Heck found that vinyl acetate underwent reaction with "phenylpalladium chloride" to give equal amounts of *trans*-stilbene, phenylacetaldehyde, and β-acetoxystyrene.[24]

† "In the synthesis of vinyl acetate, all agents necessary for its destruction are present."[287]

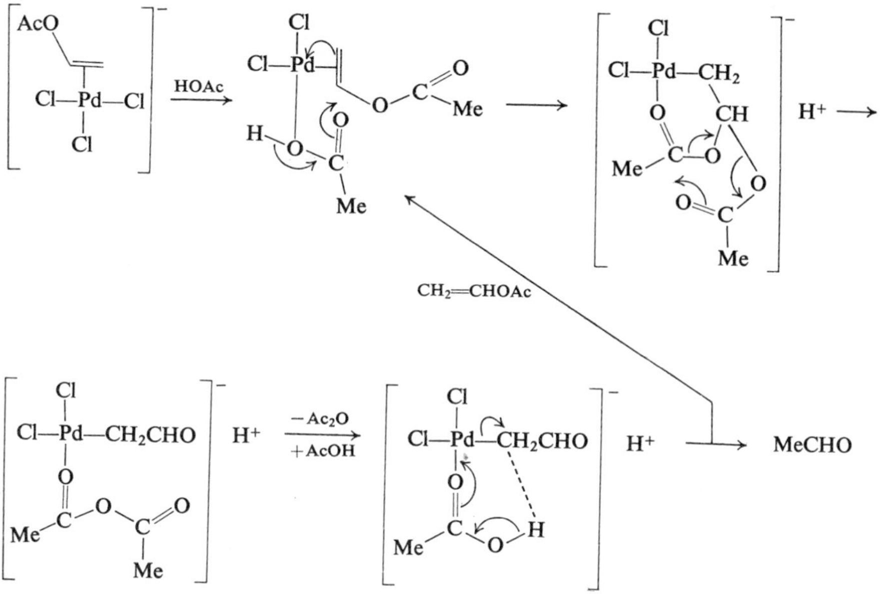

SCHEME II-2

D. THE ACETOXYLATION OF BENZENE AND ALKYLBENZENES

The acetoxylation of toluene to benzylacetate and benzylidene acetate was first reported in 1965 in a patent by Kroenig and Frenz.[369] Davidson and Triggs[223] in 1967 noted that whereas benzene reacted with pure palladium acetate to give biphenyl, in the presence of sodium acetate a 50% yield of phenyl acetate was obtained.

Both the nuclear and the side-chain oxidation have received some attention. Davidson and Triggs[435] studied the reaction of benzene and palladium acetate in acetic acid which gave biphenyl and phenyl acetate. The formation of the latter was catalyzed by sodium acetate with formation of metal, but was virtually suppressed under 60 atm of oxygen, when the main product was biphenyl. No products such as toluene or xylenes, indicating the intermediacy of acetoxy radicals (which are known to decompose to give methyl radicals and CO_2), were present in any of these reactions. This is in contrast to the reaction of benzene with lead tetraacetate which does give rise to toluene and xylenes under these conditions and which proceeds by a radical-chain reaction. However, since the formation of phenyl acetate in the palladium reaction is so strongly inhibited by oxygen, a feature characteristic of the scavenging of a free-radical intermediate, a free radical, but not a methyl or acetoxy radical, must be present.

Davidson and Triggs have rationalized this in terms of the following mechanism, where a palladium(I) species, "PdOAc," is the active oxidizing species. This arises, as mentioned before (Volume I, Chapter VI, Section C and this Volume, Chapter I, Section D) from the reaction:

$$C_6H_6 + Pd(II) \rightarrow C_6H_5Pd(II)$$

$$2C_6H_5Pd(II) \rightarrow C_6H_5 \cdot C_6H_5 + 2Pd(I)$$

Products arising from the intermediate radicals Me· and MeCOO· were also absent in the reactions of alkyl benzenes. These reactions were also strongly inhibited by oxygen and hence R can be aryl or benzyl:

$$Pd^{I}(OAc) + RH \rightarrow RH^{+} + AcO^{-} + Pd^{0}$$

$$RH^{+} \rightarrow R\cdot + H^{+}$$

$$R\cdot + Pd^{II}(OAc)_2 \rightarrow R^{+} + OAc^{-} + Pd^{I}(OAc)$$

$$R^{+} + OAc^{-} \rightarrow ROAc$$

This scheme also explains the observed inhibition by perchloric acid, since this would inhibit ionization of RH^+.

An interesting variant on this reaction by Tissue and Downs[436] described the catalytic acetoxylation and nitration of benzene by nitrite or dinitrogen tetroxide in acetic acid catalyzed by palladium acetate.

$$C_6H_6 \rightarrow C_6H_5NO_2 + C_6H_5OAc$$

These reactions may also be compared with the very curious formation of benzoic acid from benzene and acetic acid (Chapter I, Section D).

The side-chain oxidation of toluene and *p*-xylene has been investigated by Davidson and Triggs[435] and Bryant *et al.*[222, 437, 438] The initial product from toluene was benzyl acetate, which was then very slowly further acetoxylated to benzylidene acetate.

$$PhMe + OAc^{-} \xrightarrow{Pd(II)/Sn(II)/O_2} PhCH_2OAc \longrightarrow PhCH(OAc)_2$$

The reaction was strongly catalyzed by tin(II) acetate and could be carried out catalytically in acetic acid with reoxidation by air at 100°.†[437] Substituted toluenes (XC_6H_4Me, where $X = NO_2$, Cl, Ac) reacted very slowly or gave rise to side products. For a series of methylbenzenes the rates decreased in the order

† There is an apparent contradiction here to the work reported by Davidson and Triggs who found that oxygen inhibited side-chain acetoxylation.

toluene > *p*-xylene > *o*-, *m*-xylene > mesitylene > durene (1,2,4,5-tetramethylbenzene) > hexamethylbenzene, suggesting that steric effects were of great importance. These authors also observed that the rate of the reaction with toluene was independent of its concentration; this is in agreement with the Davidson–Triggs mechanism if the rate-determining step is the formation of the active palladium(I) species.

Bryant *et al.*[222] also resolved the apparent discrepancy with the work of van Helden and Verberg[219] who obtained largely (75%) bitolyls from reaction of toluene with acetate in the presence of palladium(II). They showed that, using palladium acetate–potassium acetate, the yield of phenyl acetate was very dependent on the acetate-to-palladium ratio and reached a maximum of 93% with a 20:1 ratio. The bitolyls were formed preferentially in the presence of chloride (Table II-3).

TABLE II-3

Products from Reaction of Toluene with Acetate in the Presence of Palladium(II)[a]

Catalyst	Ratio $[OAc^-]/Pd(II)$	$PhCH_2OAc$ (%)	$PhCH(OAc)_2$ (%)	Bitolyls (%)
$KOAc/Pd(OAc)_2$	5	53	6	3
	10	67	7	1
	15	69	6	—
	20	93	6	—
$NaOAc/PdCl_2$	5	2	—	64
	10	5	1	36
	15	50	5	8
	20	63	4	1

[a] After Bryant *et al.*[222]

Bryant *et al.*[438] have also reported the catalytic acetoxylation of *p*-xylene to give **(II-19)** (60–85%) and **(II-20)** (14–34%) (which may be formed synchronously) and later, at very high conversion, **(II-21)**. The rates for *o*-

$$Me{-}C_6H_4{-}Me \xrightarrow[\text{air}/OAc^-]{Pd(II)/Sn(II)}$$

$$\underset{\textbf{(II-19)}}{Me{-}C_6H_4{-}CH_2OAc} + \underset{\textbf{(II-20)}}{AcOCH_2{-}C_6H_4{-}CH_2OAc} + \underset{\textbf{(II-21)}}{Me{-}C_6H_4{-}CH(OAc)_2}$$

and *m*-xylene were approximately one-third that for *p*-xylene; similar products were obtained. See also Appendix.

Fitton *et al.*[439] have prepared the σ-benzyl complexes (**II-22**) and (**II-23**) as models for the benzylic oxidation. Both reacted with silver acetate (or, to a

$$(Ph_3P)_4Pd + PhCH_2Cl \longrightarrow$$

$$\underset{\textbf{(II-22)}}{Ph_3P(Cl)Pd(CH_2Ph)(PPh_3)} \xrightarrow[\text{hexane}]{CHCl_3} \underset{\textbf{(II-23)}}{Ph_3P(PhCH_2)Pd(\mu\text{-}Cl)_2Pd(CH_2Ph)(PPh_3)}$$

lesser extent, with potassium acetate) in acetic acid at 100° to give benzyl acetate (major product) and benzylidene diacetate. They suggested that a benzyl-palladium complex might be an intermediate in the formation of the mono- and diacetates from toluene and put forward a mechanism for this process.

Davidson and co-workers[225, 435] briefly noted that in the oxidation of hexamethylbenzene two reaction paths could be distinguished. One was the formation of pentamethylbenzyl acetate in the presence of perchloric acid and the absence of chloride ion. This was not inhibited by oxygen and therefore presumably did not arise by a free-radical mechanism. Under these conditions in the presence of oxygen duroquinone was also formed.

$$C_6Me_6 + Pd(OAc)_2 + HClO_4 \xrightarrow{40°} C_6Me_5CH_2OAc$$

$$\xrightarrow[60°\text{–}80°]{O_2\ (100\ \text{atm})} \text{duroquinone (2,3,5,6-tetramethyl-1,4-benzoquinone)}$$

A recent patent has described the direct oxidation of benzene to phenol using palladium chloride–copper(II) chloride in ethylene glycol in the presence of oxygen.[440]

More mechanistic studies are needed to clear up the anomalies presently existing in this area.

E. THE OXIDATION OF ALCOHOLS

The oxidation of alcohols by palladium salts has been known for some time.† Nikiforova *et al.*[320] observed that these reactions were much slower than those with olefins. They also described the catalytic oxidation of aqueous ethanol to acetaldehyde using $PdCl_2/CuCl_2$ and air.

Lloyd[442] found that alcohols reduced palladium(II) more readily under anhydrous than under aqueous conditions. *n*-Butanol was oxidized to a mixture of butyraldehyde and its di-*n*-butyl acetal by palladium chloride in the presence of 3 atm of oxygen. Using copper nitrate as reoxidant the acetal was the main product, together with a little butyl butyrate. More of this ester, and a lower overall yield, was obtained using copper chloride. Similar results were obtained for ethanol and other primary alcohols.‡

$$Pd(II) + RCH_2OH \rightarrow RCHO + RCH(OCH_2R)_2 + RCOOCH_2R + Pd(0)$$

$$(R = Me, Et, Pr^n, Pr^i, Ph)$$

As expected, secondary alcohols gave ketones: cyclohexanone was formed from cyclohexanol, methyl ethyl ketone from 2-butanol, etc. *t*-Butanol did not react, and may therefore be a useful solvent or cosolvent for reactions in which reduction to metal is to be avoided.

However, Ketley and Fisher noted that ethylenepalladium chloride complex reacted with *t*-butanol to give mainly *t*-butyl chloride and isobutene.[306] Isopropyl chloride was the major product from the analogous reaction with isopropanol.

The inhibition by water of the reactions of alcohols with palladium chloride can be explained in terms of the better coordinating power of water. The mechanism for the oxidation can be represented, on present evidence, as

$$RCH_2OH + (PdCl_2)_n \rightleftharpoons RCH_2O(H)\text{—}PdCl_2\text{—}(L) \xrightarrow[(L)]{-HCl}$$

$$RCH(H)\text{—}O\text{—}PdCl(L)\text{—}(L) \longrightarrow RCHO + HCl + Pd(0) + 2(L)$$

(II-24)

(L = solvent, etc.)

† Berzelius in 1828 noted that potassium tetrachloropalladate when heated in aqueous ethanol gave palladium metal.[441]

‡ Except methanol which was less easily oxidized than the others and gave largely methyl formate (HCOOMe), rather than methylal $[CH_2(OMe)_2]$.

and involves β elimination of H–Pd from the palladium(II) alkoxide intermediate (**II-24**). Very similar reactions occur between alcohols and other platinum metal halides, for example, Rh(III)[443] or Ir(III).[444] In many of *these* cases the metal hydrides are stable and can be isolated.

Davidson and co-workers[225] found that in chloride-free media, under *acidic* conditions, palladium(II) caused the oxidation of primary alcohols to acids or esters. This reaction was made catalytic by the presence of oxygen.

$$C_2H_5OH \xrightarrow{Pd(II)/H_2SO_4/H_2O/O_2/100^\circ} CH_3COOH$$

The higher oxidizing power of palladium(II) in the absence of chloride is due to the higher oxidation potential under these conditions $\{E^0[Pd^{II}/Pd^0(H_2O)]$, 0.91 V; $E^0[PdCl_4^{2-}/Pd^0(Cl^-)]$, 0.59 V$\}$.

F. THE OXIDATION OF CARBON MONOXIDE

Phillips in 1894 already showed that carbon monoxide reacted readily with an aqueous solution of palladium chloride to give metal and carbon dioxide.[157]

$$CO + PdCl_2 + H_2O \rightarrow CO_2 + Pd + 2HCl$$

The reagent, $PdCl_2/HCl$ on silica, has been proposed for the detection of subtoxic amounts of carbon monoxide; darkening is observed on exposure to the gas.[445]

The oxidation of carbon monoxide has been the subject of a number of papers by Fasman and his collaborators.[446–455]

The reaction can be carried out catalytically (in palladium) by using quinones, dichromate, copper(II), or ferric chloride as reoxidants.[446,447,455] The rate of oxidation is first-order in CO, palladium(II), and approximately so in halide concentration, the most active catalyst being $PdBr_4^{2-}$.[448–453] For other anions, the activity decreases in the order Br > Cl > NO_2 > I > SCN > CN.[451]

Fasman and co-workers also attempted to establish the stoichiometry of the reaction and of the intermediate formed. They found that carbon dioxide was formed well before metal was precipitated[448]; furthermore, three moles of CO were absorbed by the solution before one mole of CO_2 had been liberated. They suggested that two carbonyls were therefore linked to palladium in an intermediate complex, probably of palladium(0), e.g., $K_2Pd(CO)_2Br_2$. A Pd(I) or a cluster complex are alternatives; furthermore, Kingston and Scollary have shown that palladium chloride reacts with CO in protonic solvents to give hydridocarbonyl complexes such as $[PdH(CO)Cl_2]^-$.[456] The intermediate formation of a hydridocarbonyl complex under the conditions used by Fasman is therefore quite plausible.

Fasman *et al.* also suggested that formation of the carbonyl intermediate was rapid, but the subsequent oxidative hydrolysis to CO_2 and metal was much slower.[450] These reactions obviously merit further investigation.

G. THE OXIDATION OF FORMIC ACID

The oxidation of formic acid to carbon dioxide and water catalyzed by Pd(II) in acetic acid in the presence of Cu(II) has been the subject of an ICI patent[457] and a detailed study by Aguilo.[458]

$$HCOOH + [O] \rightarrow H_2O + CO_2$$

Solutions of Pd(II) in formic acid were stable, even on heating, in the absence of base (acetate, in this case); neither metallic palladium nor copper(II) had any significant effect on the decomposition. The reaction was first-order in [Pd(II)], $[OAc^-]$, and [HCOOH]; the parameters of the reaction were found to be, $\Delta H^{+} = 22$ kcal/mole and $\Delta S^{+} = -6$ cal·mole^{-1}. deg.$^{-1}$ The deuterium isotope effect, k_H/k_D, for DCOOH was 2.0. These results were interpreted in terms of the following mechanism, where the active species was the formate ion.

$$HCOOH + OAc^- \rightarrow HCOO^- + AcOH$$

$$L_4Pd(II) + HCOO^- \rightarrow HCOO\text{—}Pd(II)L_3 + L$$

$$H\text{—}C(=O)\text{—}O\text{—}PdL_3 \longrightarrow H^+ + CO_2 + Pd^0 + 3L$$

$(L = AcOH, AcO^-, \text{ or } Cl^-)$

H. MISCELLANEOUS PALLADIUM-INDUCED OXIDATIONS

An ICI patent has described the oxidation of acrolein to acrylic acid employing as catalyst, $PdCl_2/Cu(OAc)_2/LiCl/LiOAc$ in acetic acid.[459] Using the same catalyst, another ICI patent described the oxidation of allyl acetate to 3,3-diacetoxy-1-propene and acrolein. Isomerization also occurred.

$$CH_2{=}CHCHO + [O] \rightarrow CH_2{=}CHCOOH$$

$$CH_2{=}CHCH_2OAc + OAc^- + [O] \longrightarrow CH_2{=}CHCH(OAc)_2 + CH_2{=}CHCHO + MeCH{=}CHOAc$$

Heck has also described the reaction of allylic alcohols with "phenylpalladium" to give aldehydes (Chapter I, Section A).[22]

$$CH_2{=}CHCH_2OH + \text{"PhPdX"} \rightarrow PhCH_2CH_2CHO$$

Palladium complexes also oxidized isonitriles to isocyanates.

$$RNC + [O] \rightarrow RNCO$$

This reaction was first observed by Fischer and Werner.[460] More recently Hagihara and co-workers[461] have shown that isonitriles were catalytically oxidized by palladium(0) and platinum(0) complexes [e.g., $(Ph_3P)_4M$] in benzene in the presence of oxygen. The palladium(0) complex $(Ph_3P)_4Pd$ also catalyzed the oxidation of triphenylphosphine to triphenylphosphine oxide (Volume I, Chapter I, Section B,2,d).[462] The oxidation of RNC and of Ph_3P by zero-valent metal complexes, especially Ni(0), has been the subject of a detailed recent study by Otsuka *et al.*[463]

Stern[464] has reported the $(Ph_3P)_4Pd$ catalyzed autoxidation of cumene to its hydroperoxide, and has suggested a mechanism involving $(Ph_3P)_2PdO_2$ as intermediate (see also Nyman *et al.*[465]).

A last curious oxidation reaction, really involving the formation of a Si–O bond, is the following ethanolysis reaction, described by Yamamoto *et al.*[466] and which proceeds only in the presence of palladium chloride.

$$Me_3SiSiMe_2CH{=}CH_2 + EtOH + PdCl_2 \longrightarrow Me_3SiOEt + CH_2{=}CHMe_2SiOEt + EtOSiMe_2Et + Me_3SiSiMe_2Et + Pd + 2HCl$$

A similar reaction occurs for the trisilane, $Me_3SiSi(Me_2)SiMe_2CH{=}CH_2$.

I. THE OXIDATION OF PALLADIUM π COMPLEXES

Although these are not, for the most part, oxidations caused by palladium or one of its salts, reactions of some palladium complexes are summarized here for completeness. Some have already been discussed elsewhere.

1. π-Allylic Complexes

Hüttel and co-workers[333, 337] have investigated the oxidation of some π-allylic complexes with various oxidizers (see also Volume I, Chapter V, Section F,7). The simpler ones were already oxidized by palladium(II), whereas more sterically hindered π-allylic complexes needed stronger oxidizers such as dichromate or manganese dioxide. Unsaturated aldehydes were frequently obtained, but the nature of the product often depended on the pH of the reaction, for example,

$$[\text{Me-C}_3\text{H}_3(\text{Me})\text{-PdCl}]_2 \xrightarrow{PdCl_2/HOAc} \text{MeCH=CMeCHO}$$

$$\xrightarrow{PdCl_2/HOAc/OAc^-} \text{Me}_2\text{C=CHCHO}$$

$$\xrightarrow{MnO_2/H^+} \underset{(42\%)}{\text{MeCH=CMeCHO}} + \underset{(42\%)}{\text{CH}_2\text{=CMeCOMe}}$$

Attack always appeared to occur at the 1- or 3-carbons of the π-allylic group.

Green *et al.*[387] noted that chloro (π-cyclohexenyl)palladium dimer was not readily attacked by acetate ion, and therefore was not an intermediate in the oxidation of cyclohexene. However, π-allylic *acetates* easily decompose particularly in the presence of CO,[13]

$$[(\text{C}_3\text{H}_3\text{RR}')\text{-PdCl}]_2 + \text{AgOAc} + \text{CO} \rightarrow \text{RR}'\text{C=CHCH}_2\text{OAc} + \text{RR}'\text{C(OAc)CH=CH}_2$$

2. Cyclobutadiene Complexes

Maitlis and Stone[127] found that tetraphenylfuran was obtained when tetraphenylcyclobutadienepalladium chloride was decomposed by triphenylphosphine in the presence of air [in its absence, octaphenylcyclooctatetraene was formed (Volume I, Chapter IV, Section F,2)]. The reactions of 1,2-di-*t*-butyl-3,4-diphenylcyclobutadienepalladium chloride with triphenylphosphine in air gave two di-*t*-butyldiphenylfurans.[182, 183]

$$[(Ph_4C_4)PdCl_2]_2 \xrightarrow{Ph_3P/O_2} \text{tetraphenylfuran}$$

$$[(Ph_2C_4\text{-}t\text{-}Bu_2)PdCl_2]_2 \xrightarrow{Ph_3P/O_2} \text{furans (Ph, Ph, t-Bu, t-Bu)} + \text{or}$$

Nitric acid oxidation of tetraphenylcyclobutadienepalladium chloride gave *cis*-dibenzoylstilbene, which was also obtained by nitric acid oxidation of either the *endo*- or *exo*-alkoxytetraphenylcyclobutenylpalladium chloride complexes (**II-25**).[94, 179]

$$[(Ph_4C_4)PdCl_2]_2 \xrightarrow{HNO_3} PhCO(Ph)C{=}C(Ph)COPh \xleftarrow{HNO_3} [(RO\text{-}C_4Ph_4)PdCl]_2$$

(II-25)

Under the conditions of the latter reaction the cyclobutadiene complex may well be an intermediate.

Sandel and Freedman[129] showed that tetraphenylfuran is produced by autoxidation of the 1-halo-1,2,3,4-tetraphenylcyclobutenyl radical, while dibenzoylstilbene may arise from tetraphenylcyclobutadiene.

Ketones have been obtained by ethanolysis of some enyl-palladium complexes.[188, 467]

J. THE FORMATION OF C–O BONDS UNDER NONOXIDATIVE CONDITIONS

In addition to the cases where C–O bonds are formed as part of some oxidative process, a large number of reactions are also known where they arise by other means. Some have been covered in Section C and other reactions where the metal really plays a role are summarized below.

1. To Give Complexes

Alkoxy groups are incorporated into complexes in many reactions, for example, the formation of (**II-26**) from butadiene and Pd(II) in the presence of an alcohol (Volume I, Chapter V, Section F,3,b).

$$CH_2{=}CH\cdot CH{=}CH_2 + ROH + L_2PdCl_2 \longrightarrow [(\text{allyl})PdCl \; CH_2OR]_2$$

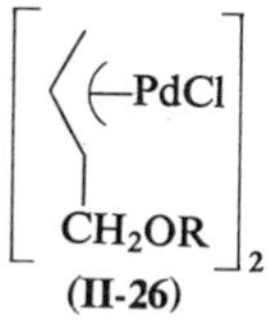

(**II-26**)

Alkoxytetraphenylcyclobutenyl complexes (**II-25**) arise from reaction of diphenylacetylene and $PdCl_2$ in alcohols (RO *endo*-), or by treatment of $[Ph_4C_4PdCl_2]_2$ with alcohols (RO *exo*-) (Volume I, Chapter IV, Section F,2, Chapter V, Section B,5, and this Volume, Chapter I, Section C,3,d). The latter is, of course, analogous to the reaction of diene–$PdCl_2$ with alcohols in the presence of base.

Carbene complexes of Pd(II) and Pt(II) are obtained on addition of alcohols to isonitrile complexes[468] or by solvolysis of cationic monoacetylene complexes of Pt(II) (Volume I, Chapter II, Section B,3).

2. To Give Organic Compounds

The dimerization of butadiene with incorporation of phenol, alcohols, or carboxylic acids has been discussed in Chapter I, Section C,3,c. The products are 1-substituted 2,7-octadienes and 3-substituted 1,7-octadienes,

$$2CH_2{=}CHCH{=}CH_2 + ROH \longrightarrow ROCH_2CH{=}CH(CH_2)_3CH{=}CH_2 + CH_2{=}CHCH(OR)(CH_2)_3CH{=}CH_2$$

(R = alkyl, acyl, aryl)

Divinylpyrans are obtained from aldehydes and butadiene.[147]

Other palladium promoted C–O bond-formation reactions include (a) the condensation of α-olefins (particularly with branched chains) and formaldehyde to 1,3-dioxanes,[198, 199] e.g.,

$$Me_2CH\cdot CH{=}CH_2 + HCHO \xrightarrow{Pd(II)/Cu(II)} \text{(1,3-dioxane)} + \text{(1,3-dioxane)}$$

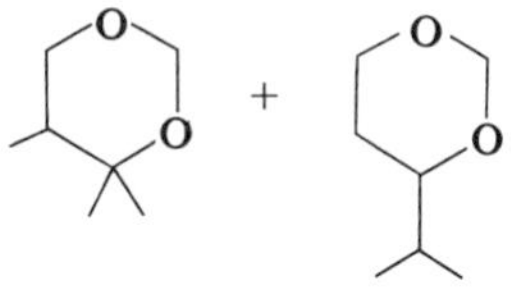

(b) the carbonylation of ethanol to diethyl carbonate (this appears to be an oxidation reaction; see also Appendix),[469]

$$\mathrm{EtOH + CO \xrightarrow{Pd(II)/Cu(II)} (EtO)_2CO + Et_2O + EtCl}$$

(c) the catalyzed addition of phosgene to ethylene oxide,[470]

$$\text{(ethylene oxide)} + \mathrm{COCl_2 \xrightarrow{PdCl_2/Al_2O_3} ClCH_2CH_2OCOCl + (ClCH_2CH_2O)_2CO}$$

and (d) the solvolysis of diketene to acetoacetates,[471,472]

$$\text{(diketene)} + \mathrm{ROH \xrightarrow{Pd(II)} MeCOCH_2COOR} \qquad (\mathrm{R = H, Et})$$

In acetone, a condensation occurred.

$$\text{(diketene)} + \mathrm{Me_2CO} \xrightarrow{Pd(II)} \text{(2,2,6-trimethyl-4H-1,3-dioxin-4-one)}$$

π-Allylic complexes were formed and may be intermediates in these reactions.

Allylic alcohols, esters and acetates undergo a variety of Pd(II) catalyzed exchange reactions[423] (see also Appendix),

$$\mathrm{CH_2{=}CHCH_2OAc + MeOH \rightarrow CH_2{=}CHCH_2OMe + AcOH}$$

These compounds also give allylamines under analogous conditions,[200,473]

$$\mathrm{CH_2{=}CHCH_2OR + Et_2NH \rightarrow CH_2{=}CHCH_2NEt_2 + ROH}$$

(R = H, alkyl, acetyl, phenyl).

Chapter III

The Formation and Cleavage of Carbon–Hydrogen Bonds

A. INTRODUCTION

Reactions in which C–H bonds are formed, and/or, are broken are very common indeed, especially in metal-catalyzed reactions. Of course, classically, the noble metals themselves have long been used for hydrogenation under heterogeneous conditions. These reactions are considered briefly in Chapter V.

Many reactions are also known which occur under homogeneous conditions in which hydrogen is added, removed, or simply moved. Some of these are incidental to other reactions which occur and have been discussed elsewhere. They include:

(1) Hydrogen shifts coincident with C–O bond formation [e.g., $CH_2{=}CH_2 \rightarrow CH_3CHO$ or $CH_3CH(OAc)_2$] (Chapter II, Section B); hydrogen shifts coincident with carbon–carbon bond formation [e.g., $CH_2{=}CH_2 \rightarrow CH_2{=}CHCH_2CH_3$; $CH_2{=}CHCH{=}CH_2 \rightarrow CH_2{=}CHCH{=}CH(CH_2)_2CH{=}CH_2$] (Chapter I, Section C).

(2) Hydrogen loss coincident with (a) carbon–oxygen bond formation [e.g., $PhMe \rightarrow PhCH_2OAc$; $RCH_2OH \rightarrow RCHO + RCH(OR')_2$] (Chapter II, Sections D, E, F, G, and H); (b) carbon–carbon bond formation (e.g.,

$C_6H_6 \rightarrow C_6H_5C_6H_5$; $PhCH{=}CH_2 + C_6H_6 \rightarrow PhCH{=}CHPh$) (Chapter I, Section D); (c) carbon–X bond formation (e.g., $C_2H_4 \rightarrow CH_2{=}CH{-}N$ [ring with C=O]) (Chapter IV).

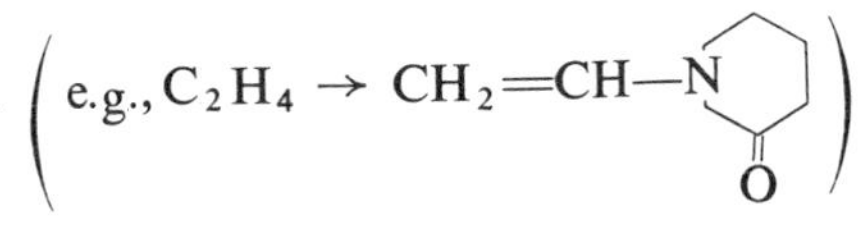

The topics dealt with in this chapter include double-bond isomerization, disproportionation, dehydrogenation, hydrogenation, and the catalyzed addition of H–X to double and triple bonds.

B. DOUBLE-BOND ISOMERIZATION

The palladium-catalyzed isomerization of double bonds was first reported in 1962 in a patent which described the palladium chloride-catalyzed isomerization of 4-methyl-1-pentene to 2-methyl-2-pentene. The carbon skeleton was not isomerized. A later patent described this in more detail, using $(PhCN)_2PdCl_2$, in the absence of solvent, as a catalyst.[474]

The palladium-catalyzed isomerizations of olefins have been the subject of considerable study, but in contrast to work on rhodium-catalyzed[475] or platinum-catalyzed[476] reactions, the results are still unclear and difficult to interpret. It may well be that the reaction paths, involving successive addition and elimination of M–H to double bonds, postulated by Cramer and Lindsey,[477] do apply here to some degree at least. However, the difficulty is that a recognizable palladium–hydride species has not been detected nor is it clear how it originates in these reactions.

The most important contributions to this field are summarized below; since the conditions (nature of substrate, catalyst and solvent) vary so greatly, the results of each group of investigators are presented separately. Where no specific solvent is mentioned, reactions were carried out in the pure olefin. The interpretations are discussed in Section B,2.

1. Experimental Results

Moiseev and his collaborators have studied the isomerization of 1-butene,[478–482] methyl-1-pentenes, and 1-hexene.[481] These reactions were usually carried out in aqueous perchloric acid in the presence of palladium chloride and chloride ion (LiCl). *Trans*- and *cis*-2-butenes were formed in the ratio of 3.2:1.†[479] The rate of isomerization was first-order in 1-butene, and inhibited by acid and chloride ion. The rate of isomerization was expressed as[478, 480]

$$\text{Rate} = \{k_1[\text{1-butene}] - k_2[\text{2-butenes}]\} \frac{[PdCl_4^{2-}]^2}{[H_3O^+][Cl^-]^2}$$

No significant incorporation of deuterium was found when reactions were carried out in D_2O rather than H_2O, but the rate of the reaction in water was 3.4 times as fast as in 85% D_2O.[482] This is close to the ratio, $k_{H_2O}/k_{D_2O} = 4.05$, observed by Moiseev *et al.*[323, 483] for the rates of oxidation of ethylene to acetaldehyde in the two solvents.

The addition of a reoxidant [copper(II) or quinone] gave rise to an induction period for the reaction. This induction period was not observed in water in the absence of reoxidant, but was observed when the reaction was run in hydrocarbons, or, to a lesser extent, in alcohol.

In a later paper, Moiseev and Grigor'ev[484] claimed that the normal stable β-$PdCl_2$ did form π-olefin complexes, but did not catalyze olefin isomerization in the absence of alcohols or water. In contrast, the metastable (at normal temperatures) α-$PdCl_2$ was more soluble and was also capable of isomerizing without the addition of water. Thus, after 1.5 hours at 60° only 3% of 1-hexene had been isomerized by β-$PdCl_2$ (to 0.8% 2-hexene and 2.2% *cis*-3-hexene), whereas 96.8% had been isomerized by α-$PdCl_2$ (to 21.6% 2-hexene, 22.3% *cis*-3-, and 52.5% *trans*-3-hexenes). These results were interpreted to mean that a reduction process, either to a hydrido or a lower valent palladium halide species, was necessary to form the true catalyst from normal β-$PdCl_2$. In the presence of oxidizers this normally fast reaction was inhibited.

Heating solid $PdCl_2$ to over 400° may be presumed to have caused the loss of some chlorine with consequent reduction of a part at least, to a lower oxidation state. It does not seem reasonable to ascribe the change in activity solely to a change in geometry of the palladium chloride (cluster → chain), although this may also be a contributory factor. However, palladium metal, even in the presence of chloride ion, does not isomerize 1-olefins.

Donati and Conti have shown that on heating to 100° with palladium chloride, 1-octene and other long-chain α-olefins gave mixtures of isomeric π-allylic complexes of two types, $[C_nH_{2n-1}PdCl]_2$ and $[(C_nH_{2n-1}PdCl)_2PdCl_2]$. These complexes, on further heating, gave conjugated dienes; for example, the π-allylic complex isomer mixture from 1-octene gave largely

† At equilibrium. However, the ratio at equilibrium also varied with the conditions and was 1.8 for $PdCl_2$ in benzene and 4–6 for $PdCl_2$ in DMF.

1,3- and 2,4-octadienes.[485,486] Isomerization only proceeded partially in DMF; the active catalyst was deactivated through the formation of $[C_nH_{2n-1}PdCl]_2$ and $(DMF_2H)_2[Pd_2Cl_6]$. Morelli *et al.*[487] further showed that the complexes $C_nH_{2n}PdCl_2DMF$ were formed initially (at $-80°$) and that these were readily transformed into the π-allylic complexes by loss of HCl.

Pregaglia *et al.*[488] found that at 5°–10° α-olefins reacted with palladium chloride in the absence of solvent to give π-olefin complexes. They further showed that for a number of 1-, *cis*-2-, and *trans*-2-olefins the uncomplexed olefin had undergone negligible isomerization when the reaction was carried at 10° or below; isomerization became significant at 25°, especially for 1-butene and 1-pentene. The degree of isomerization of the complexed olefin was also determined by liberating it with 1,5-COD. For the lower olefins, especially those prepared at 10° or below, little isomerization was again observed. This became substantial for butene complexes prepared at higher temperatures and especially for the complex derived from 1-hexene at 0°.

These workers later[107] showed that the exchange between free and complexed olefin was very fast since no separate resonances for complexed and uncomplexed olefin were visible in the NMR spectrum even at $-60°$. This led to the conclusion that isomerization occurred during complex formation or was coincident with it. The actual rates were very dependent on the olefin. Conti *et al.*[107] found that the 1-pentene *complex* was very difficult to make; most routes led to a mixture of *cis*- and *trans*-2-pentene complexes. However, it could be prepared from the γ isomer of $[cis\text{-2-pentene}PdCl_2]_2$ and 1-pentene in chloroform at 20°, and they studied its isomerization to the *cis*- and *trans*-2-pentene *complexes*. The rate of isomerization in chloroform in the presence of a narrow concentration range of cocatalyst was given by

$$\text{Rate} = k_1[\text{complex}] + k_2[\text{complex}][\text{cocatalyst}] + k_3[\text{complex}][\text{cocatalyst}]^n$$

$$(\text{complex} = [\text{1-pentenePdCl}_2]_2)$$

The first term in this rate expression was independent of cocatalyst and reflected a very slow isomerization which occurred in its absence. The cocatalyst studied most was ethanol; it was found that for a concentration range of 1 to 1.4 moles of ethanol per mole of complex, the rate was first-order in both complex and ethanol. Even in this range, however, the Arrhenius law was not obeyed. Furthermore, a concentration ratio of less than 1.0 gave anomalous results, whereas one greater than 1.4 gave a considerable acceleration in rate, coincident with precipitation of metal. The effect of the alcohol was thought to be due to the breaking of a chlorine bridge only since other alcohols, olefins (2-hexene), and pyridine were also effective (to varying degrees). Overall the kinetics were interpreted in terms of a fast preequilibrium to give **(III-1)** followed by two different isomerization paths to give the *cis*- and *trans*-2-pentene complexes **(III-2)** and **(III-3)** ($k_{cis}/k_{trans} = 1.6$). The isomerization **(III-2)** ⇌ **(III-3)** was very much slower in both directions and showed an induction period. The latter observation was interpreted in terms of the primary isomerization to a 1-pentene complex.

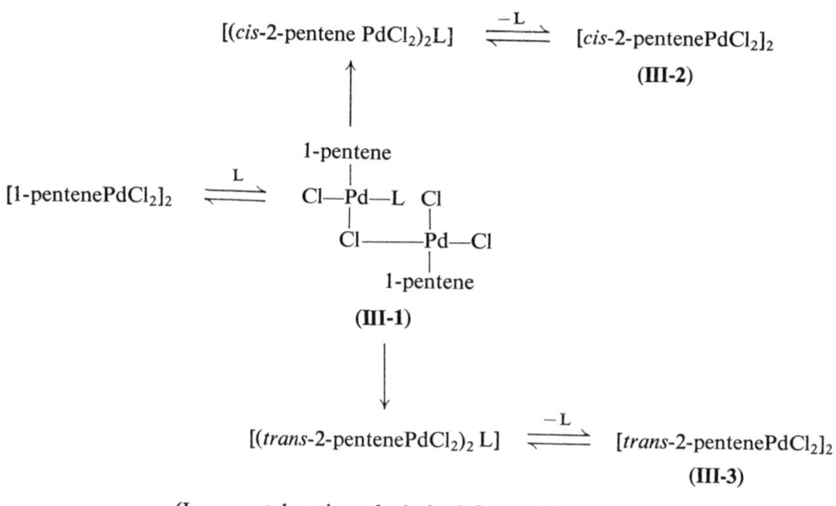

(L = cocatalyst, i.e., alcohol, olefin, pyridine, etc.)

Bond and Hellier[489] studied the isomerization of pentenes in benzene, catalyzed by $(PhCN)_2PdCl_2$, and again found that 1-pentene isomerized faster than *cis*- or *trans*-2-pentene. Furthermore, an induction period, the length of which varied inversely with catalyst concentration, was observed for the isomerization of *cis*- to *trans*-2-pentene. This induction period was eliminated on addition of 5% 1-pentene, a result which again suggested that here, at any rate, cis–trans isomerization proceeded via isomerization to the 1-alkene. Furthermore, the catalytically active species appeared to be formed more readily from 1-pentene. Again, in the isomerization of 1-pentene, the cis:trans ratio was at first higher than the equilibrium value, but as the reaction proceeded, it approached this value. The rate of isomerization of *cis*-2-pentene was first-order in olefin after the initial induction period. Bond and Hellier also observed that the isomerization rates in ethyl acetate or methyl ethyl ketone were approximately the same as in benzene. In alcohols, metal was precipitated.

A more detailed study of the isomerization of 4-methylpentenes was published in 1965 by Sparke *et al.*[490] The catalyst employed was $(PhCN)_2PdCl_2$, but the ethylene- or cyclohexenepalladium chloride complexes functioned equally well. Palladium chloride gave erratic results owing to solubility problems, and $(PhCN)_2PdBr_2$ was too insoluble. Olefin–palladium chloride complexes were considered to be the true catalysts. The relative rates of isomerization of the isomers were

(98) > (34) > (25) > (8)

1-Butene isomerized more slowly than the methylpentenes. In all cases isomerization again proceeded from terminal to internal double bonds. Here again, the isomerization of 4-methyl-1-pentene gave more *cis*-4-methyl-2-pentene than the equilibrium value. Most of their results were interpretable in terms of a stepwise migration of the double bond, but Sparke *et al.* did note that in the isomerization of 4-methyl-1-pentene, 2-methyl-1-pentene was initially formed to a greater extent than apparently allowed by this mechanism.

Asinger *et al.*[491] investigated the isomerization of 1-octene and tritium-labeled $CH_2{=}CHCHT(CH_2)_4CH_3$. 1-Octene was again found to isomerize stereoselectively at low conversions; palladium chloride was a much better catalyst than "rhodium trichloride hydrate" or platinum(II) chloride. At 80°, "activated" palladium chloride† caused 96 % isomerization of 1-octene; in an exchange experiment with 1-hexene under these conditions 49 % exchange of tritium with the hexene was observed. At 25° the degree of isomerization and exchange dropped to 47 and 8 %, respectively. However, when acetic acid (in a 1:1 molar ratio to palladium) was used as cocatalyst virtually no exchange of tritium between the octene and the acetic acid was detected.

The lack of exchange between 1-octene and acetic acid (this time, CH_3COOD) in the presence of palladium chloride had already been noted earlier by Davies.[492,493] Davies also examined the isomerization of $CH_2{=}CHCD_2C_5H_{11}$ and was not able to detect any significant amount of deuterium at the terminal carbon after isomerization had occurred. This led him to conclude that only 1,2-hydrogen shifts and no 1,3-hydrogen shifts took place. Cruikshank and Davies[494] studied the isomerization of allylbenzene to *trans*-propenylbenzene **(III-4)** (95 %) and *cis*-propenylbenzene **(III-5)** (5 %) by $Na_2Pd_2Cl_6$ in acetic acid at 65°. The rate was found to be independent of the olefin concentration and proportional to that of the catalyst; however, *trans*-propenylbenzene appeared to cocatalyze the isomerization of allylbenzene. An induction period was again observed for the reaction without cocatalyst.

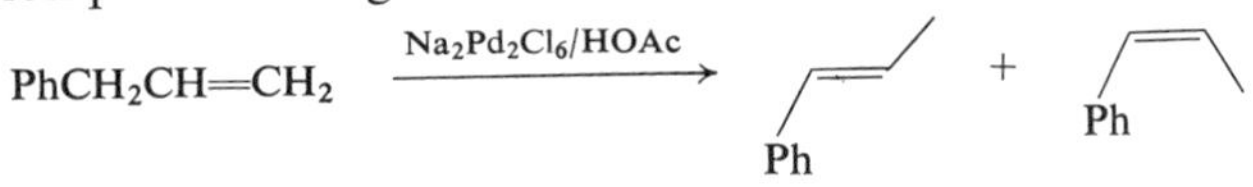

$$PhCH_2CH{=}CH_2 \xrightarrow{Na_2Pd_2Cl_6/HOAc} \text{(III-4)} + \text{(III-5)}$$

Ph Ph

(III-4) **(III-5)**

† This was activated by dissolving in hydrochloric acid and evaporating the solution to dryness. It was then considerably more soluble.

The importance of steric factors in the isomerization was shown by the fact that 2-bromo-3-phenyl-1-propene did not isomerize under these conditions and that 2-methyl-3-phenyl-1-propene isomerized very slowly to β,β'-dimethylstyrene.

$$PhCH_2\overset{\overset{Br}{|}}{C}{=}CH_2 \rightarrow \text{no reaction}$$

$$PhCH_2\overset{\overset{Me}{|}}{C}{=}CH_2 \rightarrow PhCH{=}CMe_2$$

Davies *et al.*[495] also studied the $(PhCN)_2PdCl_2$-catalyzed (no solvent) isomerization of 4-phenyl-1-butene **(III-6)**.

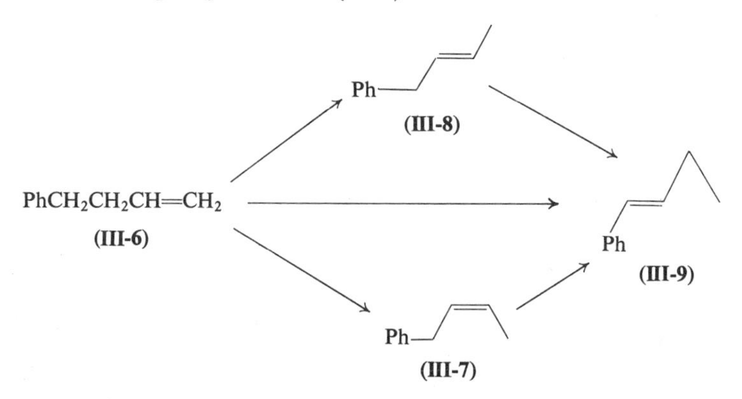

The isomerization paths found are shown above. Again *cis*-4-phenyl-2-butene **(III-7)** was formed in preference to the trans isomer **(III-8)**. Both the cis and trans isomers isomerized at about the same rate, approximately one-twelfth that for **(III-6)**.

An induction period was again observed when **(III-6)** was isomerized; the kinetics of the reaction after the induction period were interpreted as being first-order in catalyst and in olefin. Again, more of the cis isomer **(III-7)** than the equilibrium value was found early in the reaction. In addition, the amount of the final product, **(III-9)**, obtained was greater than could be accounted for on the basis of two stepwise paths, **(III-6)** → **(III-7)** → **(III-9)** and **(III-6)** → **(III-8)** → **(III-9)**. Hence some isomerization **(III-6)** → **(III-9)** *effectively* occurred directly.

The reverse reaction did not appear important, nor was any *cis*-1-phenyl-1-butene detected in the products.

The isomerization of 1-hexene by $(PhCN)_2PdCl_2$ in benzene at 60°–90° was briefly reported by Harrod and Chalk.[496] Under these conditions, and especially in the presence of air, moisture, or cocatalysts such as ethanol, rapid precipitation of metal occurred with consequent complete loss of catalytic activity.† They observed that, in the absence of cocatalyst, the palladium(II)-catalyzed reaction was about as rapid as catalysis by $[ethylenePtCl_2]_2$ with ethanol as cocatalyst, and gave similar isomer distributions. In contrast to other workers, they found that the palladium-catalyzed reaction needed no cocatalyst.

In a later study, these authors[498] noted that in the isomerization of 1-olefins by $(PhCN)_2PdCl_2$ in benzene at 55° all possible internal olefins were formed at roughly comparable rates. This was in contrast to the "rhodium trichloride hydrate"-catalyzed reactions which proceeded in a recognizably stepwise manner. They also observed that deuterium transfer from 1-heptene-3-d_2 to unlabeled pentene occurred during isomerization, but to a lesser extent than for rhodium(III) catalysts. Harrod and Chalk also attempted to measure the extent of deuterium shift in $C_3H_7CD{=}CDH$ and in $C_4H_9CD_2CH{=}CH_2$ during isomerization. The results, while not conclusive, appeared to show that in $C_3H_7CD{=}CDH$ little movement of deuterium occurred. This was consistent with 1,3-hydrogen shifts:

$$CH_3CH_2CH_2CD{=}CDH \rightarrow CH_3CH_2CH{=}CDCDH_2 \rightarrow CH_3CH{=}CHCHDCDH_2$$

In the second case, in contrast to the work of Davies,[492, 493] migration of deuterium from the allylic carbon onto both terminal and nonterminal vinylic carbons occurred. Harrod and Chalk accounted for the latter again in terms of a 1,3-hydrogen shift.

$$C_3H_7CH_2CD_2CH{=}CH_2 \rightarrow C_3H_7CH_2CD{=}CHCH_2D \rightarrow C_3H_7CH{=}CDCH_2CH_2D$$

However, the considerably greater enrichment observed at the terminal vinylic carbon and other anomalies could not be explained.

Cramer and Lindsey[477] investigated the isomerization of 1-butene, using a different catalyst, Li_2PdCl_4, in the absence of solvent. In this case trifluoroacetic acid was found to be a useful cocatalyst, allowing the isomerization to proceed significantly even at 30°. Using labeled trifluoroacetic acid (CF_3COOD or CF_3COOT) a small amount of the label was found to be incorporated into the butenes after 30% isomerization had occurred. The catalyst Li_2PdCl_4–

† Metallic palladium also isomerizes alkenes, but at a much higher temperature. For example, Carrà and Ragaini showed that 1-butene was isomerized by palladium on alumina in the gas phase at 200°.[497]

CF_3COOH also interconverted *cis*- and *trans*-2-butenes. These authors estimated the relative rates of the various isomerization processes as:

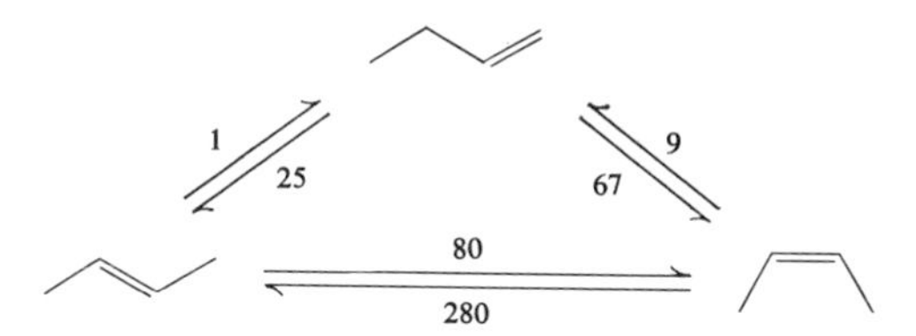

As indicated, *cis*-2-butene was formed preferentially on isomerization of 1-butene and reached its equilibrium value only toward the end of the reaction.

For Li_2PdCl_4 the isomerization was cocatalyzed by strong acids (e.g., $HClO_4$), but inhibited by chloride ion.

1-Butene was also isomerized rapidly in methanol in the presence of hydrogen; in aqueous media metal was precipitated. Stannous chloride deactivated the catalyst. Using H_2–Li_2PdCl_4 in MeOD, some incorporation of deuterium into the olefins was noted.

In experiments with allylically labeled $CH_3CD_2CH{=}CH_2$ Cramer and Lindsey found that (a) deuterium was not lost during isomerization, (b) undeuterated 1-butene was isomerized more slowly than the deuterated material, and (c) the deuterium was largely present in 2-butene as $CH_3CD{=}CDCH_3$.

Volger,[215] in some related studies, has shown that 2-methylallylpalladium chloride derived from isobutene and palladium chloride in CH_3COOD was not labeled unless the system had been allowed to equilibrate for some time before sodium acetate was added. In that case deuterium was found at C-1 (C-3) of the π-allylic complex.

$$Me_2C{=}CH_2 + PdCl_4^{2-} \overset{A}{\rightleftharpoons} (Me_2C{=}CH_2)PdCl_3^- + Cl^- \overset{B}{\rightleftharpoons} \frac{1}{2}\left[Me\text{-}(\pi\text{-}C_3H_4)PdCl\right]_2 + HCl + 2Cl^-$$

A series of equilibria of the above type were postulated. Deuterium exchange can only occur via equilibrium B. Since, in the absence of base, this is well over to the left, and in the presence of base, well over to the right, exchange only occurs very slowly. Volger also noted that on treatment of a π-allylic complex with DCl a monodeuterated olefin resulted.

2. Summary and Conclusions

As can be seen, the situation is extremely confused and the problem needs a great deal more careful attention before any firm mechanistic conclusions are drawn. One problem which has bedeviled some work in this field has been olefin purity; small traces of impurities, particularly peroxides, have led to statements[493,496] about rates of isomerization which were later retracted.[495,498] Similar, as yet unrecognized difficulties, may explain some of the other inconsistencies noted above. It would appear that most progress toward the determination of mechanisms is to be made in actually determining the rates of isomerization of the olefin complexes themselves as has been initiated by Conti *et al.*[107] This approach, where feasible, would avoid unknown factors such as the equilibria:

$$\text{Olefin} + \text{Pd(II)} \rightleftharpoons \text{olefinPd(II)}$$

The conclusions listed below are very tentative for the most part.

(1) Palladium(II) is a very effective catalyst for isomerization, particularly of terminal to internal olefins. Skeletal rearrangements do not occur to any detectable extent. Palladium chloride itself is not a very effective catalyst and gives erratic results owing to solubility problems. Better catalysts are $(PhCN)_2PdCl_2$ or $PdCl_4^{2-}$. It appears to be more effective than most other transition metal complexes though few comparative studies have been made. In contrast, metallic palladium, sometimes precipitated from these reactions, is virtually inactive. A further point on which there appears to be general agreement is that chloride (or halide) is a necessary cocatalyst; a number of workers have noted that olefin isomerization is largely repressed in chloride-free media.

(2) The rate of isomerization appears to be pseudo first-order in the olefin; large amounts of both acid and chloride ion (as in the Wacker process) inhibit the reaction. The order in palladium(II) complex is uncertain; both first- and second-order dependence have been observed.

(3) In aqueous media the reaction is slower in D_2O than in H_2O; furthermore, deuterated olefins appear to isomerize more slowly than undeuterated ones.

(4) Many workers have observed induction periods in their studies. These could be due to (a) the presence of oxidizing agents or peroxide impurities; (b) the active catalyst in the system, e.g., a palladium complex in a lower oxidation state, being formed slowly; or (c) isomerization not proceeding directly, but rather via an intermediate which is formed slowly (e.g., *cis*- $\rightleftarrows$

trans-2-pentene isomerization appears to proceed via 1-pentene). A combination of some or all of these effects is also possible.

(5) Cocatalysts have been used (a) to overcome the above induction period and (b) to increase the rate of isomerization. Cocatalysts of the first type include ethanol and other alcohols, as well as 1-alkenes. The function of the alcohols is probably to break a PdClPd bridge, that of the latter is not so clear. A number of workers have found that small amounts of the acids of anions which complex weakly ($HClO_4$, CF_3COOH) will greatly accelerate rates of isomerization, particularly cis $\rightleftharpoons$ trans.

(6) Exchange of label (either deuterium or tritium) occurs between two different olefins under isomerization conditions. It is not as extensive as for rhodium-catalyzed reactions, and the label also appears to be distributed rather differently in the products. On the other hand, exchange between olefin and cocatalyst (CH_3COOD, CF_3COOD, CF_3COOT, or CH_3OD) is much less, of the order of 0.5% or so. This may still be quite significant.

(7) It appears that longer chain olefins isomerize more rapidly. The rate is also dependent on the degree of branching and on the size of substituents on the olefin. Large substituents (Br) are apparently able to repress isomerization entirely.

(8) In the isomerization of 1-alkenes, the *cis*-2-alkene is always formed preferentially at first. Later, when the system has had time to establish equilibrium, the thermodynamically expected cis : trans 2-alkene ratio is established.

In contrast to the stepwise nature of olefin isomerization catalyzed by other metals, palladium(II) catalysts appear to form all the possible double-bond isomers right from the start, and at roughly comparable rates.

In general, 1-alkenes are isomerized faster than internal alkenes, but Cramer and Lindsey[477] have noted that cis–trans isomerization of 2-butenes is very much faster than other isomerizations, using a Li_2PdCl_4–CF_3COOH catalyst. However, this result may be anomalous and reflect an acid-catalyzed rather than a metal-catalyzed reaction.

(9) Three experiments designed to investigate the movement of deuterium during isomerization gave two totally different results for allylically labeled olefins $RCD_2CH{=}CH_2$. Davies[492,493] and Cramer and Lindsey[477] interpreted their results in terms of a deuterium shift from C-3 to C-2, whereas Chalk and Harrod[498] interpreted theirs in terms of a C-3 to C-1 shift. On the other hand, deuterium in $C_3H_7CD{=}CDH$ did not appear to move during isomerization.[498] It is obvious that this area will also require closer investigation using more sophisticated techniques than have been used to date.

In view of all this, it is too early to consider mechanistic paths too seriously. The following suggestions have been put forward.

The *metal–hydride, metal–olefin hydride, metal–alkyl* path involves a chain reaction, initiated by Pd–H species. Considerable speculation about the origin of the hydride and the nature of the species has been made. Definite evidence is unfortunately lacking; two possibilities are that the hydride arises from the cocatalyst or the olefin.[477, 499, 500]

$$-Pd-H + RCH_2CH{=}CH_2 \rightleftharpoons \underset{RCH_2CH \stackrel{\perp}{=} CH_2}{-Pd-H} \rightleftharpoons RCH_2CH_2CH_2-Pd-$$

$$\rightleftharpoons \underset{RCH_2CHCH_3}{-Pd-}$$

$$-Pd-H + RCH{=}CHCH_3 \rightleftharpoons \underset{RCH \stackrel{\perp}{=} CHCH_3}{H-Pd-}$$

Evidence in favor of this hypothesis includes the observed transfer of deuterium (or tritium) between two different olefins. A hypothesis of this type would also explain the observed induction periods, the adverse effects of oxidizing agents, and part, at least, of the effect of cocatalysts. It also implies that hydrogen (or deuterium) moves in a series of 1,2- rather than 1,3-shifts.

The second popular hypothesis involves an *intermediate π-allylic palladium hydride complex.*[498, 500]

$$-Pd^{II}- + RCH_2CH{=}CH_2 \rightleftharpoons \underset{RCH_2CH \stackrel{\perp}{=} CH_2}{-Pd^{II}-} \rightleftharpoons \underset{\textbf{(III-10)}}{H{>}Pd^{IV}\text{—}(CH_2{-}CH{-}CH{-}R)}$$

$$-Pd^{II}- + RCH{=}CHCH_3 \rightleftharpoons \underset{RCH \stackrel{\perp}{=} CHCH_3}{-Pd^{II}-}$$

In this scheme no outside source of hydride is needed, and yet a hydridic intermediate, (**III-10**), is present. Intramolecular hydrogen (deuterium) transfer occurs by a 1,3-shift. In principle, it could also allow hydrogen (deuterium) transfer between two olefins; however, this would probably be less facile than with hydridic intermediates of the type previously discussed.

The difficulties with this approach are that **(III-10)** is then effectively a palladium(IV) complex, and neither the induction period, nor the effect of cocatalyst observed is readily understandable.

Moiseev and Grigor'ev have endeavored to overcome these objections by postulating that the reactive species (under their conditions, in water) is not a palladium(II) complex but a dimeric complex, such as $Pd_2Cl_4^{2-}$, effectively of Pd(I). One metal atom then can coordinate the olefin, and form the π-allylic complex, while the other abstracts the hydrogen,† e.g.,[484]

$$Pd_2{}^{I}Cl_4^{2-} + RCH_2CH{=}CH_2 \rightleftharpoons \left[\begin{array}{c} R(H)C{-}CH{=}CH_2 \\ Cl{-}Pd^{I}{-}Pd^{I}{-}Cl \\ Cl \quad Cl \end{array}\right]^{2-}$$

$$\rightleftharpoons \left[\begin{array}{c} H \quad CH_2{\cdots}R \\ Cl{-}Pd^{II} \quad Pd^{II}{-}Cl \\ Cl \quad Cl \end{array}\right]^{2-} \rightleftharpoons \left[\begin{array}{c} CH_3CH{=}CHR \\ Cl{-}Pd^{I}{-}Pd^{I}{-}Cl \\ Cl \quad Cl \end{array}\right]^{2-}$$

$$\rightleftharpoons Pd_2Cl_4^{2-} + RCH{=}CHCH_3$$

The advantages of this reaction scheme and similar ones based on a Pd(I) intermediate are that it accounts for the induction period and the lack of reaction under oxidizing conditions, as well as for the observation by Moiseev *et al.* that the isomerization was second-order in palladium complex. The necessity for invoking the participation of Pd(IV) is also avoided. However, it would appear to need an appreciable lifetime for the Pd(I) species and Davidson and Triggs have shown that Pd(I) is quite unstable in the presence of chloride ion and disproportionates rapidly to Pd(II) and Pd(0). Furthermore, this scheme would probably be applicable only to reactions carried out in fairly polar solvents and could not explain the isomerizations described by Conti *et al.*, which apparently take place inside Pd(II) complexes.

† Author's note: I have taken some liberties with the scheme as originally proposed and this is not an exact representation of it.

Alper *et al.*[501] have recently shown that 1,4-cyclohexadiene forms 1,3-cyclohexadieneiron tricarbonyl with $Fe(CO)_5$ by a 1,3-hydride shift via a π-allylic hydride complex. Boennemann[502] reported a direct observation of the equilibrium,

$$\text{(}\pi\text{-allyl)}-Ni^{II}(PF_3)(H) \rightleftharpoons (\text{MeCH}=\text{CH}_2)-Ni^{0}-PF_3$$

but insufficient detail was given. These experiments suggest another possibility, namely that in the palladium catalyzed reactions, the active species is a Pd(0) complex which can give a π-allylic palladium(II) hydride but which, under adverse conditions, is deactivated by the formation of metal.

The possibility that a 1,3-hydrogen shift may not involve a discrete metal hydride intermediate and yet be dependent on a direct interaction between the metal and the migrating hydrogen has been discussed by Cowherd and von Rosenberg in connection with the iron pentacarbonyl-catalyzed isomerization of allylic alcohols to ketones.[503] Palladium(II) can catalyze analogous reactions, for example,[504]

$$\underset{\text{HO}\ \ \text{OH}}{\text{PhCHCHCH}=\text{CH}_2} \xrightarrow[\text{DMF}/150^\circ]{(Ph_3P)_2PdCl_2} \underset{\text{HO}\ \ \text{O}}{\text{PhCHCC}_2\text{H}_5}$$

and it is possible that a similar mechanism may apply here too.

In connection with their study of the isomerization of 4-phenyl-1-butene, Davies *et al.*[495] have suggested an explanation for the approximately simultaneous appearance of all the isomers and have reconciled this with a stepwise 1,2-hydrogen shift mechanism. This scheme assumes that the isomerizations occur in the complexes and that, in going from isomer A to isomer C, the reaction goes via palladium(II) olefin complexes of A, B, and C, written as Pd–A, Pd–B, and Pd–C.

If the rate of isomerization of Pd–B to Pd–C is faster than the displacement of B from Pd–B by another olefin, then Pd–B is, but olefin B itself is not, an intermediate. The reaction still proceeds in a stepwise manner, but C can be produced even faster, under suitable circumstances, than B.

$$\begin{array}{ccccc}
\text{A} + -\text{Pd}- & & \text{B} + -\text{Pd}- & & \text{C} + -\text{Pd}- \\
\updownarrows & & \updownarrows & & \updownarrows \\
\text{A}-\text{Pd}- & \rightleftharpoons & \text{B}-\text{Pd}- & \rightleftharpoons & \text{C}-\text{Pd}-
\end{array}$$

The extent of this type of reaction will be governed by the rates at which the olefins are liberated from their complexes. Little is known about this.

As Conti *et al.*[107] have found that exchange of pentenes with pentene–palladium chloride π complexes is very fast even at $-60°$, this implies that the isomerization of olefins is a very fast reaction indeed, if it goes by a 1,2-hydride shift.

At present an explanation of this type seems most satisfactory. The assumption is then necessary that the active intermediate is a hydridopalladium chloride complex, which is present in very low but kinetically significant amounts during the reaction. The induction period then arises from the necessity to form the hydride, and the inhibition by oxidizing agents is readily understood. It would also explain the necessity for a cocatalyst when the reaction is run under anhydrous conditions which do not allow a Wacker-type reaction to occur.

$$RCH{=}CH_2 + PdCl_2 + H_2O \rightarrow RCOCH_3 + HCl + \text{“HPdCl”}$$

The very small degree of incorporation of label from deuterium- or tritium-labeled catalyst agrees with the inability of workers to detect a hydrido intermediate spectroscopically, and implies that each molecule of catalyst is able to cause many isomerizations. This hypothesis is also in broad agreement with the results of Cramer and Lindsey using labeled olefins.

It is also found that halide ion is necessary for isomerization to occur to a significant extent. The species “HPdCl” presumably has, therefore, a much longer lifetime than “HPdOAc.” The nature of the active species is not known, but one possibility would be a solvated $HPdCl(PdCl_2)_x$.

If the concept that Pd(II) is able to cause very fast 1,2-hydride shifts is accepted, then it becomes logical that a similar process occurs during the Wacker reaction where it is known that all the hydrogens in the product arise from olefinic hydrogens (Chapter I, Section B,2). In this case the intermediate, which must have a very short lifetime by comparison with the time needed for olefin exchange to occur, is a vinyl alcohol complex (**III-11**). The

$$HOCH(H)-CH_2-Pd^{II}- \rightleftharpoons CH_2{=}CH(OH)\cdots Pd(H) \rightleftharpoons$$

(**III-11**)

$$CH_3-CH(O-H)-Pd- \longrightarrow CH_3CHO + Pd^0 + H^+$$

complex (**III-11**) arises from the β-hydroxyethylpalladium complex by β elimination; Pd–H then readds to the complexed olefin in the reverse sense to give the α-hydroxyethylpalladium complex which then decomposes irreversibly to products. This can be envisaged either as shown, or to involve another β elimition of H from the hydroxyl group.

Other reactions, for example, the formation of ethylidene diacetate and of acetals from ethylene (Chapter II, Sections B,4 and 6) in which similar shifts occur can be explained in this manner too.

Two palladium hydride complexes have been reported (Volume I, Chapter II, Section D), but the properties so far investigated do not appear very relevant to the problems of hydrogen transfer. It would, in any case, be unrealistic to expect very similar behavior from a hydrido complex specifically made to be kinetically inert in order that it could be isolated.

Other double-bond isomerizations in which palladium(II) complexes act as catalysts include of 1,5- to 1,4 and 1,3-cyclooctadiene,[71,77,505–507] cyclopropane to propylene,[92] and phenylcyclopropane to *trans*-propenylbenzene.[356] The isomerization of 1-butene on a solid catalyst (palladium acetylacetonate on silica gel) has been reported by Misono.[508]

C. MISCELLANEOUS ISOMERIZATION REACTIONS

A large number of other isomerization reactions are also known in which hydrogen transfer can occur or where it is not clear whether it occurs since routes not involving hydrogen transfer are possible. These have largely been dealt with elsewhere and are merely summarized below.

(1) In the formation of π-allylic complexes from cyclopropanes, for example,[509] (Volume 1, Chapter V, Section B,3)

$$\text{(vinylcyclopropane)} + [C_2H_4PdCl_2]_2 \longrightarrow [\text{(}\pi\text{-allyl)PdCl (with Cl-substituted side chain)}]_2 + [\text{(}\pi\text{-allyl)PdCl (}CH_2CH_2Cl\text{)}]_2$$

Davidson *et al.*[225] found that 1-hexene gave two different π-allylic complexes with palladium acetate; the expected 1-propyl-π-allyl complex was formed under neutral conditions and the 1-ethyl-3-methyl-π-allyl complex in the presence of acid (Volume 1, Chapter V, Section B,1).

(2) A number of diolefins form π-diene complexes with isomerization (Chapter I, Section F,4).[257, 510]

(3) The complex $[Cl(Me_2C_2)_3PdCl]_2$ has been shown by Maitlis and co-workers[188] to decompose to vinylpentamethylcyclopentadiene on reaction with triphenylphosphine and similar ligands at low temperatures. The mechanism has been elucidated and involves a 1,2-hydride shift (Chapter I, Section C,3,d).

(4) Many examples of cis–trans isomerization are also known, for example, of propenyl acetates[427] (Chapter II, Section C,1); methyl oleate to methyl elaidate[511]; π-allylic complexes[512] (Volume I, Chapter V, Section F); and of dimethyl *cis,cis*-muconate to the trans,trans isomer.[84]

(5) Hydrogen transfer reactions must also occur during the formation of 1,4-disubstituted *trans,trans*-2,3-diphenylbutadienes from diphenylacetylene and various olefins[123] (Chapter I, Section C,2).

D. DISPROPORTIONATION

Hüttel and co-workers in 1961 found that cyclohexene was disproportionated to cyclohexane and benzene (3:1) by palladium chloride in acetic acid at 20°.[214, 333] Small yields of complexes and cyclohexanone were also obtained. Similar reactions have been reported by Karol and Carrick in a patent[513] and by Odaira *et al.*[33] The latter workers noted that cyclohexene underwent cyanation by palladium cyanide in acetonitrile. The products were 2-cyclohexenenitrile (9 %), cyclohexanenitrile (6 %), benzene (6 %), and cyclohexane (7 %).

$$\text{cyclohexene} + Pd(CN)_2 \xrightarrow{MeCN} \text{2-cyclohexenenitrile (CN)} + \text{cyclohexanenitrile (CN)} + \text{cyclohexane} + \text{benzene}$$

In the absence of solvent the cyanation was repressed and benzene and cyclohexane were formed catalytically.

The disproportionation has long been known with heterogeneous catalysts, and was originally demonstrated by Zelinsky and Pavlov.[514] These authors also found that both 1,3- and 1,4-cyclohexadienes underwent a very facile palladium-catalyzed disproportionation to cyclohexane and benzene. In a more recent study of the disproportionation of cyclohexene, Carrà and Ragaini[515, 516] came to the conclusion that this reaction proceeded via primary formation of 1,3-cyclohexadiene.

Since the reaction is catalyzed so strongly even by traces of metal, it is not yet clear whether it can also occur under homogeneous conditions.

The disproportionation of vinylcyclohexene to ethylcyclohexane (33 %) and ethylbenzene (67 %) by palladium has also been reported.[513]

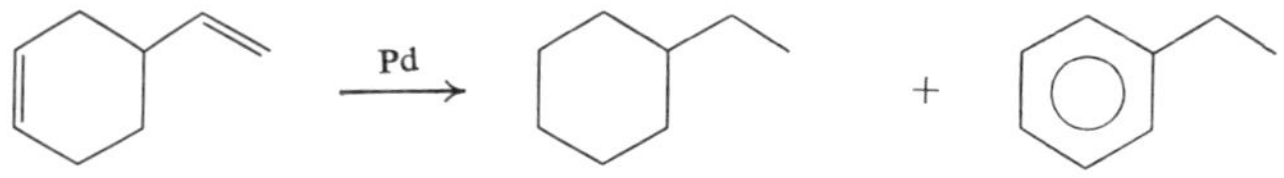

Pd

+

Disproportionation reactions under homogeneous conditions may also occur. For example, a reinvestigation by Schenach and Caserio[517] of the hydroxide-promoted cleavage of π-allylic complexes to olefins (Volume I, Chapter V, Section F,6) has shown that negligible incorporation of deuterium into the olefin occurred in D_2O.

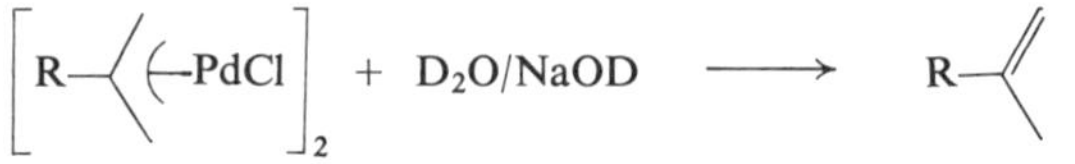

$[R-C_3H_4-PdCl]_2$ + $D_2O/NaOD$ ⟶ R–

(R = H; 83 % d_0, 13 % d_1)
(R = Me; 93 % d_0, 7 % d_1)

These authors suggested that a hydrogen was abstracted from a neighboring π-allylic group to give a fragment RC_3H_3 which then added oxygen or water to give the observed oxidized by-products. Yields of the olefin were 58 % for R = H and 63 % for R = Me.

E. DEHYDROGENATION

The thermal decomposition of some larger π-allylic complexes leads to the formation of a diene, metal, and HCl.†[486, 518] In the steroid field, Howsam and McQuillin[518] have used this reaction to effect a dehydrogenation:

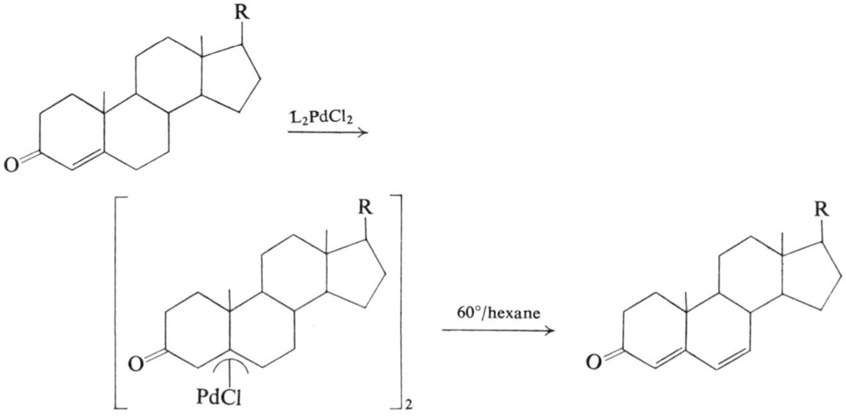

R

L_2PdCl_2

O

R

O

PdCl

2

60°/hexane

R

O

† The ease of this decomposition definitely appears to be related to the size of the ligand.

Cyanide or malonate ion can cause similar decomposition of steroid π-allylic complexes.[519]

Davidson and his co-workers[225,390] observed the formation of benzene during the oxidation of cyclohexene with palladium acetate in acetic acid. In the absence of oxygen and in the presence of perchloric acid only trace amounts were observed, but benzene became the major product under neutral or basic (sodium acetate) conditions (see Chapter I, Section B,5).

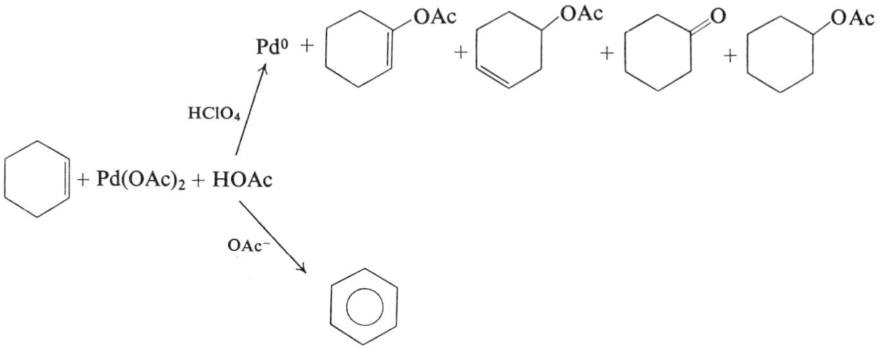

Pd^0 + OAc + OAc + O + OAc

$HClO_4$

+ $Pd(OAc)_2$ + HOAc

OAc^-

1,3-Cyclohexadiene and 1,4-cyclohexadiene were also readily dehydrogenated to benzene by palladium acetate under basic conditions; 1,4-cyclohexadiene underwent this reaction even in the presence of oxygen. A possible route for the formation of benzene is outlined below.

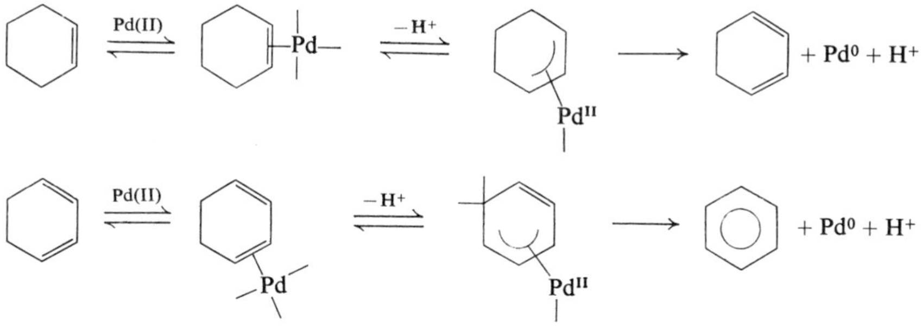

Pd(II) Pd— $-H^+$ Pd^{II} + Pd^0 + H^+

Pd(II) Pd $-H^+$ Pd^{II} + Pd^0 + H^+

F. HYDROGENATION

1. Organic Compounds

While the use of metallic palladium (particularly when supported on carbon) as a hydrogenation catalyst is very well established (Chapter V, Section A), much less attention has been paid to catalysis by homogeneous palladium catalysts.

The use of palladium chloride as a catalyst in the borohydride reduction of cyanides, nitro, and nitroso compounds to amines was reported by Pesez and Burtin.[520]

$$PhCN \rightarrow PhCH_2NH_2$$

Palladium chloride was reported by Rylander *et al.*[521] to have a very high activity in the catalytic hydrogenation (H_2) of dicyclopentadiene under homogeneous conditions in DMF or DMAC. Thiophene accelerated the rate of hydrogenation.

Maxted and Ismail[522] found that palladium chloride functioned particularly well as a catalyst in the hydrogenation of ethyl crotonate under aqueous conditions in the presence of sodium acetate.

The use of dichlorobis(tributylphosphine)palladium as a hydrogenation cocatalyst (in the presence of triiobutylaluminum, or borohydride) for 1-alkenes has been mentioned.[523,524] The analogous nickel complex showed much higher reactivity.

An ICI patent describes the use of the mixed catalyst $(Bu_3P)_2PdPtCl_4$ for homogeneous hydrogenation of terminal acetylenes to 1-alkenes.[525]

The only serious study so far of palladium(II) complexes as catalysts for homogeneous hydrogenation has been by Itatani and Bailar.[511,526] They found that dichlorobis(triphenylphosphine)palladium, particularly in the presence of stannous chloride, was an active catalyst for the hydrogenation of long-chain polyunsaturated esters. The chief features of this catalyst system were (1) isomerization of *cis* to *trans* double bonds, (2) the migration of isolated double bonds to give conjugated dienes, and (3) the selective hydrogenation of polyolefins to monoolefins only. No reduction to saturated esters occurred. Ester exchange also took place when the reaction was run in alcohols. The order of decreasing activity of the catalysts was $(Ph_3P)_2PdCl_2 + SnCl_2 > (Ph_3P)_2PdCl_2 + GeCl_2 > (Ph_3P)_2Pd(CN)_2 > (Ph_3As)_2Pd(CN)_2 > (Ph_3P)_2PdCl_2 \gg (Ph_3As)_2PdCl_2$. Slow hydrogenation also occurred with $(Ph_3P)_2PdCl_2 + SnCl_2$ in the absence of hydrogen in methanol solution.

Bailar and his collaborators also studied analogous reactions using nickel and platinum catalysts of the general type $(R_3E)_2MX_2$ (R = aryl, alkyl; E = P, As, Sb; M = Ni, Pt; X = halogen, SCN, CN).†[526,528–531] There appeared to be a close similarity between the reactions catalyzed by the palladium and the platinum complexes. The nickel complexes, however, behaved somewhat

† In the absence of the stabilizing ligand, R_3E, deposition of metal occurred during hydrogenation.[527]

differently; no cocatalyst such as stannous chloride was necessary† and hydrogenation was faster than isomerization.[528] For the platinum-catalyzed reaction, particularly using $(Ph_3P)_2PtCl_2$–$SnCl_2$, the course of the reaction depended rather on the structure of the olefin (diene) used.[530, 531] In some cases at least, isomerization and migration of double bonds (of dienes) to give conjugated dienes preceded hydrogenation, which then stopped at the mono-olefin stage.[526] In this case a mechanism, involving a labile metal hydride intermediate $[(Ph_3P)_2Pt(H)SnCl_3]$, which added to and eliminated from the unconjugated diene, was postulated to explain the isomerization to the conjugated diene. The hydrogenation of the conjugated diene was considered to arise by a binuclear allylic complex (**III-12**). Some support for the formulation (**III-12**) comes from the recently determined structure of $[C_3H_5PtCl]_4$, which has σ and π bonded allyls bridging pairs of metal atoms.[531a] However, further evidence is still needed.

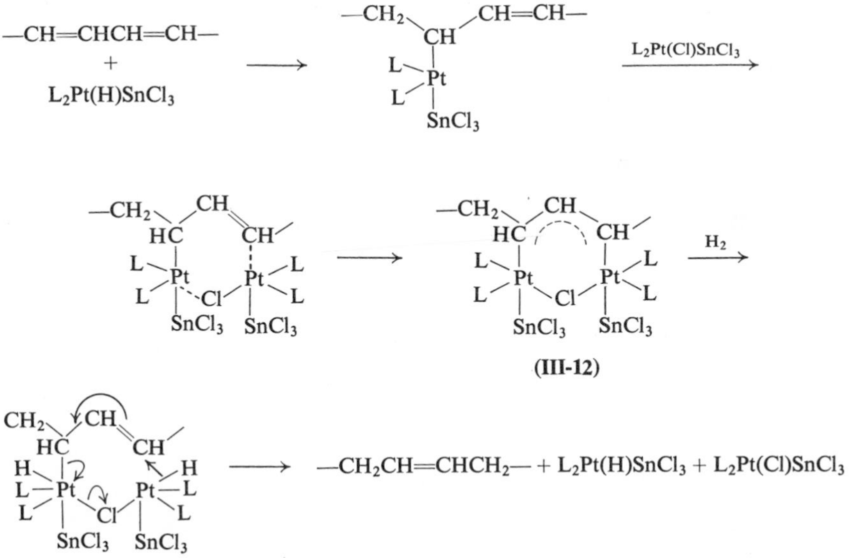

$$\text{—CH=CHCH=CH—} + L_2Pt(H)SnCl_3 \longrightarrow \text{—CH}_2\text{—CH(CH=CH—)—}PtL_2(SnCl_3) \xrightarrow{L_2Pt(Cl)SnCl_3}$$

$$\text{—CH}_2\text{—HC(—CH=CH—)—}[L_2Pt(SnCl_3)(\mu\text{-}Cl)PtL_2(SnCl_3)] \longrightarrow \text{(III-12)} \xrightarrow{H_2}$$

$$\longrightarrow \text{—CH}_2\text{CH=CHCH}_2\text{—} + L_2Pt(H)SnCl_3 + L_2Pt(Cl)SnCl_3$$

Giustiniani *et al.*[531b] made the interesting observation that, although neither $(Et_3P)_2PtHCl$, nor $(Et_3P)_2PtH_2Cl_2$ were active in the hydrogenation of olefins, the former, together with hydrogen chloride, hydrogenated olefins in a stoichiometric reaction for which the following mechanism was proposed:

† The role of the stannous chloride was to labilize the trans ligand, X, in $(R_3P)_2M(X)SnCl_3$. It is interesting that in the palladium complexes not bearing phosphines or similar ligands, addition of stannous chloride resulted in inhibition of activity.

$$L_2PtHCl + \gt C{=}C\lt \rightleftharpoons L_2Pt(H)Cl(\gt C{=}C\lt) \rightleftharpoons L_2PtCl(\gt C{-}C(H)\lt) \underset{}{\overset{HCl}{\rightleftharpoons}} L_2Pt(H)Cl_2(\gt C{-}C{-}H) \longrightarrow L_2PtCl_2 + \gt CH{-}CH\lt$$

A recent patent[531c] has described the hydrogenolysis of benzyl alcohols to toluenes, catalyzed by palladium chloride.

2. Palladium Complexes

Hydrogenation, or reduction, under suitable conditions has long been a method for determining the structures of organic ligands in complexes. However, as rearrangements often occur, the information obtained is of rather limited value unless the mechanism is known.

Christ and Hüttel found that π-allylic complexes, when treated with methanolic potassium hydroxide, were smoothly degraded to a monoolefin and metal[532] (see Volume I, Chapter V, Section F,6 and Section D, this chapter). Hydrogen in the presence of a catalyst, borohydride, and lithium aluminum hydride have all been used as to reduce ligands and sometimes give different types of products as the reactions of (**III-13**) indicate[533] (see also Volume I, Chapter V, Section F,9).

$$[\text{Ph}{-}C_3(\text{Ph})(\text{Cl})(\text{Ph})(\text{H}){-}\text{PdCl}]_2 \; (\textbf{III-13}) \xrightarrow{KOH/MeOH} PhCH{=}CPhCH_2Ph$$

$$(\textbf{III-13}) \xrightarrow{H_2} PhCCl{=}CPhCH_2Ph$$

$$(\textbf{III-13}) \xrightarrow{LiAlH_4} PhCH_2CHPhCH_2Ph$$

G. ADDITION OF H–X TO OLEFINS

A last class of reactions in which a carbon–hydrogen bond is effectively formed is that in which HX (where X is not hydrogen) is added to an olefin. A variety of such reactions are known, most of which have been discussed elsewhere.

One example of such a reaction is the palladium-catalyzed dimerization of butadiene; in the presence of alcohols and similar compounds with active hydrogen the 1-substituted 2,7-octadienes were obtained[10,11,134,137,139] (Chapter I, Section C,3,c). Takahashi *et al.* have shown, using CH_3OD, that the deuterium appeared largely in the 6-position.[139]

$$2CH_2{=}CHCH{=}CH_2 + MeOD \xrightarrow{(Ph_3P)_2Pd\text{—(maleic anhydride)}} MeOCH_2CH{=}CH(CH_2)_2CHDCH{=}CH_2$$

A (π-allylic)palladium complex has been suggested as an intermediate and Medema *et al.* showed that bis(π-allyl)palladium gave propylene and allyl methyl ether with methanol.[10]

$$(\pi\text{-}C_3H_5)_2Pd + MeOH \longrightarrow CH_2{=}CHMe + CH_2{=}CHCH_2OMe$$

A further reaction, of a somewhat different type, also catalyzed by a palladium(0) complex, is the addition of H–CN to olefins, discussed on p. 17.

Although no reports have so far appeared, it would seem likely that addition of H–X is a general reaction for olefins in the presence of Pd(0) complexes. Acids (HCl, $HOOCCF_3$) react with acetyleneplatinum(0) complexes to give, first, σ-vinylic complexes and then the free olefins and a Pt(II) complex (Volume I, Chapter III, Section E,4).

Chapter IV

The Formation of Bonds between Carbon and Halogen, Nitrogen, Sulfur, or Silicon

In the previous three chapters the formation of carbon–carbon, carbon–oxygen, and carbon–hydrogen bonds, under the influence of a soluble palladium compound have been discussed. In addition, bonds from carbon to other nonmetals, in particular, C–Cl and C–N, arise in the course of some reactions.

Comparatively little is known about such reactions, but they appear quite widespread. Some have already been mentioned incidental to other reactions. These and others are described here.

A. CARBON–HALOGEN BONDS

These reactions largely involve formation of C–Cl bonds; few reports of the formation of C–F, C–Br, and C–I bonds have yet appeared.

1. Formation of C–Cl Bonds in Complexes

A number of reactions are known in which C–Cl bonds arise concurrently with the formation of π-allylic complexes.

1,3-Dienes react with palladium chloride to give 1-(chloromethyl)allyl-palladium chloride complexes. This transformation was first observed by Shaw,[535,536] who demonstrated that butadiene reacted with bis(benzonitrile)-palladium chloride to give **(IV-1)**.

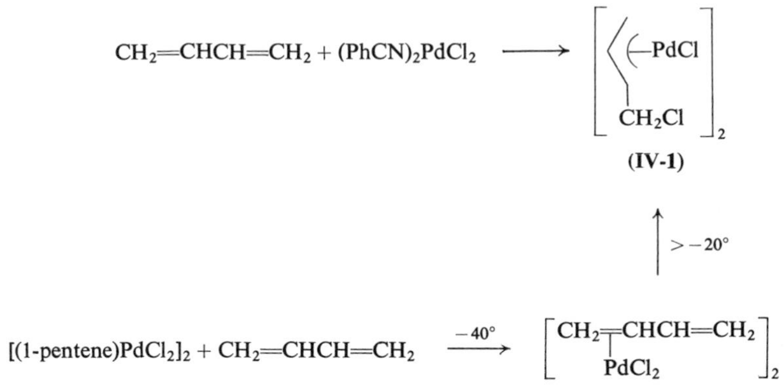

$$CH_2{=}CHCH{=}CH_2 + (PhCN)_2PdCl_2 \longrightarrow [(\pi\text{-}C_3H_4CH_2Cl)PdCl]_2$$

(IV-1)

$\uparrow > -20°$

$$[(1\text{-pentene})PdCl_2]_2 + CH_2{=}CHCH{=}CH_2 \xrightarrow{-40°} [CH_2{=}CHCH{=}CH_2 \cdot PdCl_2]_2$$

(IV-2)

A later study by Donati and Conti[537] showed that a butadienepalladium chloride complex, **(IV-2)**, was formed at low temperature (–40°) from a soluble olefin–palladium complex and butadiene. Above –20° this new olefin complex was transformed to **(IV-2)** (Volume I, Chapter V, Section B,2,a).

Rather similar reactions, in which a Pd–Cl bond has effectively added to a C=C bond have been reported for vinylcyclopropanes,[509,538] 1,2-dienes,[539–542] and a cyclopropene[533] (Volume I, Chapter V, Sections B,2,b and B,3). No intermediates were isolated in the last two reactions. However, Lupin *et al.* showed that on treatment with an excess of a phosphine, allene was regenerated from **(IV-3)**.[542]

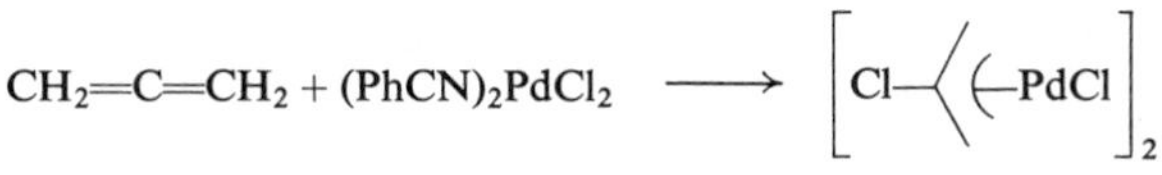

$$CH_2{=}C{=}CH_2 + (PhCN)_2PdCl_2 \longrightarrow [(2\text{-Cl-}\pi\text{-}C_3H_4)PdCl]_2$$

(IV-3)

$$\textbf{(IV-3)} + PR_3 \rightarrow (R_3P)_2PdCl_2 + CH_2{=}C{=}CH_2$$

A cis-insertion is therefore a very plausible reaction path.

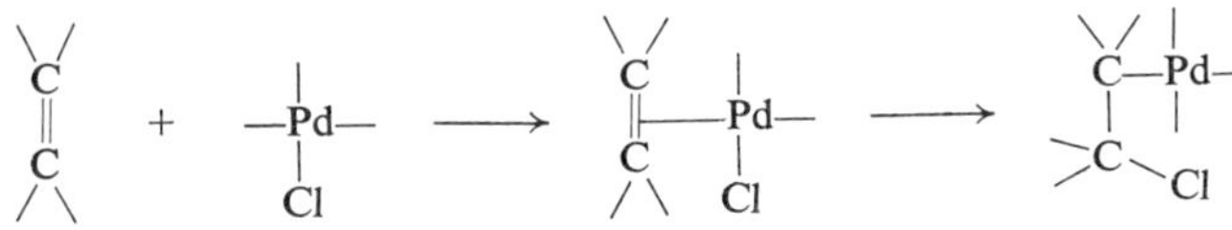

Since simple olefins normally show little tendency to undergo this reaction, it appears that the olefin has to be activated by a second double bond, cyclopropane, etc., before reaction can occur.

A rather different type of reaction is provided by the reactions of palladium chloride with acetylenes. Yukawa and Tsutsumi[543] found that reaction between palladium chloride and some 3-dimethylamino-1-propynes in the presence of excess chloride ion in methanol gave the cyclic complexes **(IV-4)**.

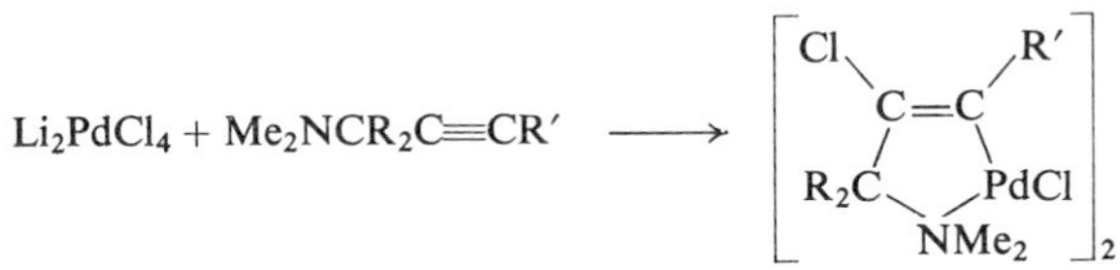

$$Li_2PdCl_4 + Me_2NCR_2C{\equiv}CR' \longrightarrow$$

(IV-4)

(R = H, R′ = Ph; R = Me, R′ = H, D)

In contrast to the above reactions where the direction of the addition of Pd–Cl is not known, but is probably cis since they were carried out in relatively nonpolar solvents, here the Pd . . . Cl has effectively added trans to the triple bond. It is certainly possible that attack by Cl^- rather than a concerted addition is occurring.

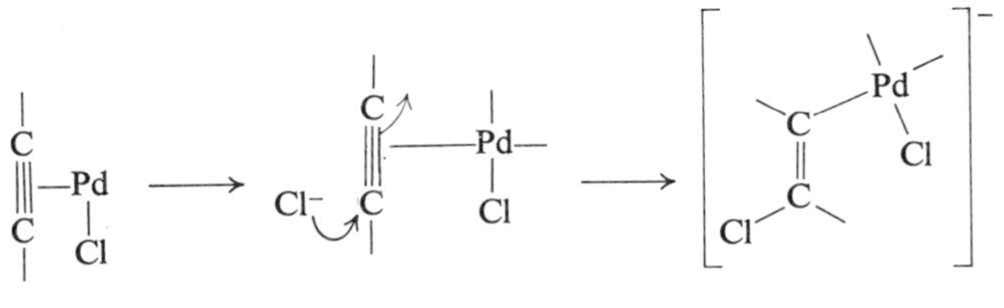

Again here, in contrast to complexes such as **(IV-1)** where the chlorine is very labile and easily replaced (e.g., by alkoxy, Volume I, Chapter V, Section F,3,b) the chlorine in **(IV-4)** is not very reactive. Silver acetate only replaced the palladium-bonded chlorine.[543]

Maitlis and co-workers[186, 187] have postulated that a cis addition of Pd–Cl to a triple bond occurs as the first step in the trimerization of 2-butyne. Subsequent steps insert further acetylenes into the palladium–vinyl bond to give, finally, in this case, **(IV-5)** (R = Me) (Chapter I, Section C,3,d).

Products which are possibly analogous to **(IV-5)** but which have not been adequately characterized have been obtained by Brailovskii *et al.*[544] and Avram *et al.*[190, 192]

2. Formation of C–Cl Bonds Resulting from Decomposition of Complexes

Hüttel and Kratzer in 1959, noted that some of the simpler π-allylic complexes broke down thermally to the metal and the allylic chloride.[545]

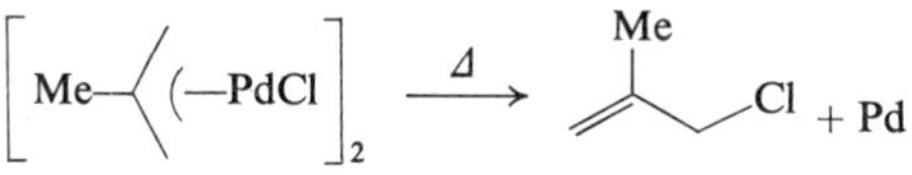

$$[\text{Me–}(\text{allyl})\text{–PdCl}]_2 \xrightarrow{\Delta} CH_2{=}C(Me)CH_2Cl + Pd$$

This reaction has been studied thermogravimetrically by Zaitsev *et al.*[546] However, longer chain π-allylic complexes lost HCl and the metal and gave a diene (Volume I, Chapter V, Section F,5).

Blomquist and Maitlis[175] and later workers[131, 179] found that tetraphenylcyclobutadienepalladium chloride decomposed thermally to 1,4-dichlorotetraphenylbutadienes (Volume I, Chapter IV, Section F,2).

An unusual reaction in which a C–Cl bond was formed was the decomposition of the complex **(IV-5)** (R = Me) with triphenylphosphine, triphenylarsine, or triphenylstibine at elevated temperatures. A major product was the (α-chlorovinyl)pentamethylcyclopentadiene **(IV-6)**.[188] The mechanism of this reaction has been discussed in Chapter I, Section C,3,d.

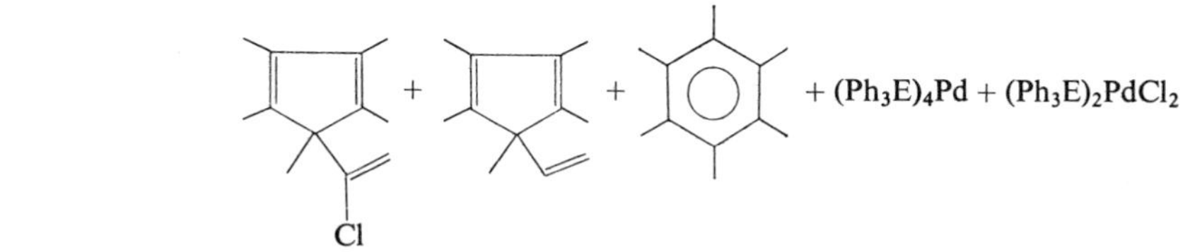

$$\textbf{(IV-5)} \xrightarrow{Ph_3E/60^\circ} \textbf{(IV-6)} + \ldots + \ldots + (Ph_3E)_4Pd + (Ph_3E)_2PdCl_2$$

(IV-5)

Cl

(IV-6)

(E = P, As, Sb)

Tsuji and his co-workers have shown that π-olefinic and π-allylic complexes react with carbon monoxide under mild conditions to give halogen-containing organic compounds.[48,49,52,61,81]

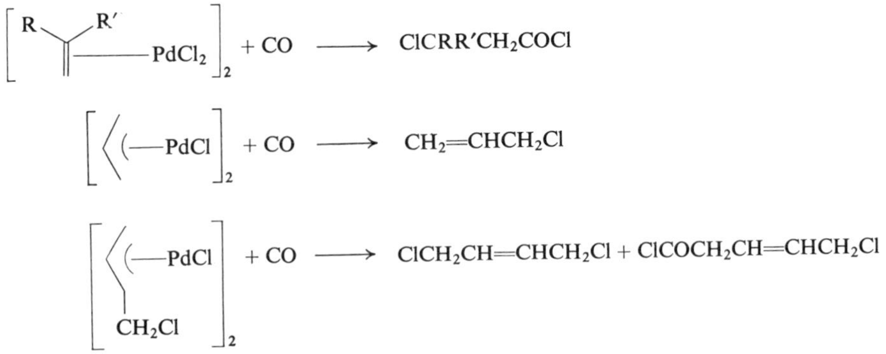

$$[\text{RR'C=CH}_2\text{—PdCl}_2]_2 + \text{CO} \longrightarrow \text{ClCRR'CH}_2\text{COCl}$$

$$[(\pi\text{-C}_3\text{H}_5)\text{PdCl}]_2 + \text{CO} \longrightarrow \text{CH}_2\text{=CHCH}_2\text{Cl}$$

$$[(\pi\text{-ClCH}_2\text{C}_3\text{H}_4)\text{PdCl}]_2 + \text{CO} \longrightarrow \text{ClCH}_2\text{CH=CHCH}_2\text{Cl} + \text{ClCOCH}_2\text{CH=CHCH}_2\text{Cl}$$

Similar reactions have been described by Medema *et al.*[11]; they were also carried out without isolation of the complex.[62–64] In particular, allylic chlorides were converted catalytically into allylic acid chlorides using allylpalladium chloride as catalyst[11,50,56,58] (Chapter I, Section F,11).

$$\text{RCH=CH}_2 + \text{CO} + \text{PdCl}_2 \longrightarrow \text{ClCHRCH}_2\text{COCl} \xrightarrow{\text{R'OH}} \text{ClCHRCH}_2\text{COOR'}$$

$$\text{CH}_2\text{=CHCH}_2\text{Cl} + \text{CO} \xrightarrow{(\text{C}_3\text{H}_5\text{PdCl})_2} \text{CH}_2\text{=CHCH}_2\text{COOCl} \xrightarrow{\text{R'OH}} \text{CH}_2\text{=CHCH}_2\text{COOR'}$$

Cyclopropane reacted analogously.[92]

3. Formation of an Organic Chloride Induced by Palladium

In addition to the above reactions, organic halides are frequently obtained as by-products. The most important are the chloro alcohols, chloroacetates, and similar compounds isolated in oxidation reactions of ethylene (Chapter II, Sections B,1 and B,4).

van Gemert and Wilkinson[104] noted that 2-chlorobutane and 1,2-dichloroethylene were major by-products from the decomposition of ethylenepalladium chloride dimer in dioxane at 50°. Similar side reactions, for example, the formation of cyclohexyl chloride from cyclohexene, have been noted by Tsuji *et al.*[72]

However, Tayim[547] has very recently reported that high yields of vinyl chloride can be obtained directly from ethylene and $PdCl_2$ if very polar solvents such as formamide are used. The reaction is made catalytic by addition of chloroanil. In less polar solvents dimerization to butenes occurred.

$$C_2H_4 + PdCl_2 \xrightarrow[80°/20\ Atm.]{HCONH_2} CH_2{=}CHCl$$

Chloro alcohols,[279] mono- and dichloroacetaldehyde,[548] and α-chloroketones[348, 549] are significant by-products of the Wacker acetaldehyde (and methyl ethyl ketone) syntheses. A patent[550] has described the copper-catalyzed α chlorination of aldehydes and ketones; these products may also arise in this way in the Wacker reaction:

$$CH_3COR + CuCl_2 + H_2O \rightarrow ClCH_2COR + Cl_2CHCOR + CuCl$$

In acetic acid chloroacetates sometimes become the major product[300, 388, 408–411] for example;

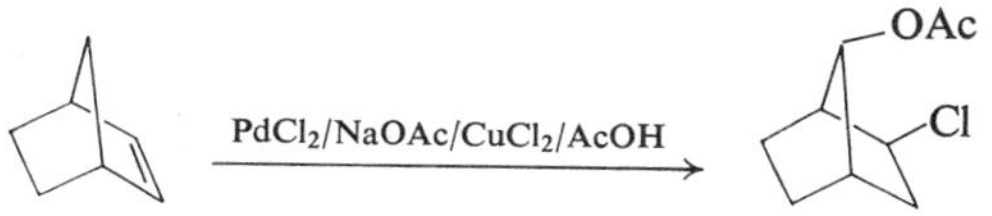

(50–84%)

Even when not the major product, a chloroacetate can be a troublesome by-product in reactions of ethylene with acetate. Henry[300] Clark *et al.*,[361, 363] and van Helden *et al.*[285] studied the formation of β-chloroethyl acetate and found that, as expected, the yield increased with increase in chloride ion concentration. Henry found that this product was only formed in the presence of copper(II) as reoxidant, and Heck[20] has shown that benzylmercury(II) compounds reacted very readily with copper chloride to give the chloride, even in the absence of Pd(II).

$$PhCH_2HgCl + CuCl_2 \rightarrow PhCH_2Cl + CuCl + Hg$$

In addition, patents have reported both ethyl and vinyl chloride as by-products in the Wacker reaction.[270, 551]

Heck[20] has described the formation of β-phenylethyl chlorides from "phenylpalladium" and olefins in the presence of copper chloride,

$$PhHgCl + Li_2PdCl_4 + CuCl_2 + RCH{=}CH_2 \xrightarrow{HOAc} PhCH_2\underset{\displaystyle R}{\underset{|}{C}}H{\cdot}Cl$$

and Henry has obtained benzoyl chloride by carbonylation of "phenyl-palladium."[45]

Alkyl chlorides also result as products from the reaction of palladium chloride, especially in the presence of copper chloride, on alcohols.[305,469]

It appears likely that C–Cl bonds result very often from a C–Pd σ-bonded intermediate in the presence of copper(II) chloride. A transmetalation to give a C–Cu–Cl intermediate is certainly a possibility here.

However, organic chlorides also arise from reactions in the absence of copper. A good example of this is the Pd(II)-catalyzed ortho chlorination of azobenzene, reported by Fahey.[552]

$$PhN{=}NPh + Cl_2 + PdCl_2 \xrightarrow{\text{dioxane–}H_2O}$$

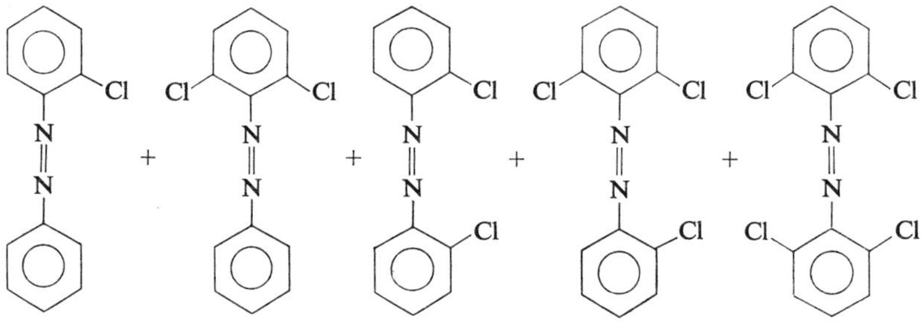

(12%) + (22%) + (30%) + (33%) + (3%)

This presumably involves cleavage of the Pd–C σ bond in (**IV-7**) and similar complexes which arise by electrophilic palladation of the aromatic ring (Volume I, Chapter II, Section C,3,b,v and this Volume, Chapter I, Section D). Intermediates isolated in this reaction were (**IV-8**) and (**IV-9**) in a 1:4 ratio.

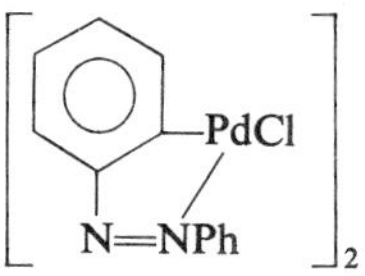

(**IV-7**) 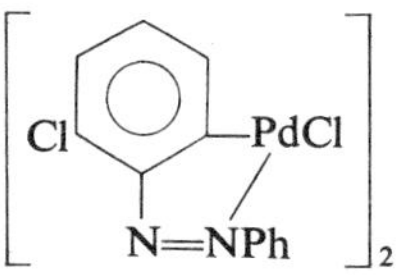 (**IV-8**) 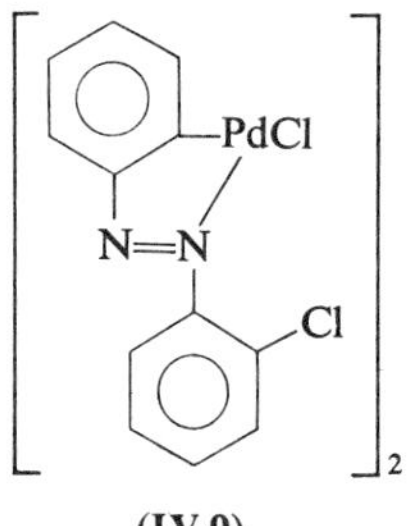 (**IV-9**)

Patents have described the chlorination of hydrocarbons using LiCl–$PdCl_2$,[553] and the carbonylation of aryl chlorides and bromides to the aroyl halides.[47]

4. Cleavage of C–Cl Bonds

Reactions in which C–Cl bonds are broken include the synthesis of π-allylic complexes from allylic halides,[56, 554] the reaction of vinyl chloride with acetate to give vinyl acetate,[420, 422, 426, 429] and the formation of allylic esters from allylic chlorides[431] (Volume I, Chapter V, Section B,3 and this Volume, Chapter II, Section C,2).

5. Formation of C–F, C–Br, and C–I Bonds

Patents to Tamura and Yasui[424] list a number of routes to vinyl fluoride.

$$CH_2{=}CHY + Pd(OAc)_2 + LiF \rightarrow CH_2{=}CHF$$
$$(Y = Cl, OAc)$$

Lupin and Shaw[541] described the bromo analog of (**IV-3**) and Maitlis and Stone[204] obtained methyl iodide from reaction of dimethylbipyridylpalladium with iodine.

B. CARBON–NITROGEN BONDS

Quite a wide variety of reactions, based upon the nucleophilicity of primary (and secondary) amines, and analogous to these of water, alcohols, and carboxylic acids (Chapter II, Section B) with olefins are known. However, they have not yet been individually explored as fully as those involving C–O bond formation.

Paiaro and co-workers[555, 556] showed that ammonia and primary and secondary amines reacted with diene complexes of palladium(II), and particularly platinum(II), to give complexes such as (**IV-10**), which then lost HCl to give the amine-bridged dimers (**IV-11**). The isomers of (**IV-10**) and (**IV-11**)

$$\text{(diene)}PtCl_2 \xrightarrow{R_2NH} \text{(IV-10)}\ [R_2\overset{+}{N}H\text{–}CH_2\text{–C}_8\text{ ring}]\overset{-}{Pt}Cl_2 \xrightarrow{-HCl} [R_2N\text{–}CH_2\text{–C}_8\text{ ring}\ PtCl]_2\ \text{(IV-11)}$$

(**IV-10**) (**IV-11**)

were also obtained.

Similar reactions have also been described by Tada *et al.*[467] and others.[183,557] Tada *et al.* showed that 1,5-cyclooctadienepalladium chloride gave an amino-cyclooctenyl complex **(IV-12)** with ammonia, which was reduced by boro-hydride to give a 60% yield of cyclooctylamine. Sodium azide gave a complex

$$\text{(COD)PdCl}_2 \xrightarrow{NH_3/C_6H_6} [\text{(NH}_2\text{-C}_8\text{H}_{12})\text{PdCl}]_2 \xrightarrow{BH_4^-} \text{C}_8\text{H}_{15}NH_2$$

(IV-12)

analogous to **(IV-12)** but the C–N_3 linkage did not survive further reactions.[467]

Panunzi *et al.*[269] have shown that attack by diethylamine on an optically active 1-buteneplatinum(II) complex is exo and therefore quite analogous to the reactions of diene complexes with amines.

$$\text{(Et-CH=CH}_2\text{)Pt((}S\text{)-NH}_2\text{CHMePh)Cl}_2 \xrightarrow{Et_2NH} \text{Et}_2\text{NH}^+\text{-CH(Et)-CH}_2\text{-Pt}^-\text{((}S\text{)-NH}_2\text{CHMePh)Cl}_2 \xrightarrow{H^+} (S)\text{-CH}_3\text{-C(H)(Et)-Et}_2\text{NH}^+$$

It is reasonable to suppose, therefore, that the first step in the reactions of olefinpalladium(II) complexes with amines is a similar one.

The vinylation of amines and amides has also been reported.[304,312,420,555,558]

$$C_2H_4 + HNRR' \rightarrow CH_2{=}CHNRR'$$

This reaction occurred in the presence of palladium chloride and base (phosphate, etc.) and could, when the substrate was an amide, be run catalytically in the presence of copper(II) acetate.[558] Since the enamines resulting from the vinylation of amines were highly reactive and difficult to purify and characterize, they were usually hydrogenated to the alkylamine.[304,420]

Hirai *et al.*[559] have also reported the use of olefin *complexes* in these reactions,

$$\left[\text{C}_6\text{H}_{10}\text{—PdCl}_2\right]_2 + \text{BuNH}_2 \longrightarrow \text{C}_6\text{H}_{10}\text{=NBu}$$

$$[\text{C}_2\text{H}_4\text{PdCl}_2]_2 + \text{(CH}_2)_3\text{C(=O)NH} \longrightarrow \text{(CH}_2)_3\text{C(=O)NCH=CH}_2$$

Another variant of the general reactions of olefins with nucleophiles catalyzed by Pd(II) is the formation of organic isocyanates from 1-octene in sulfolane in the presence of potassium cyanate. This is described in an ICI patent,[560] but no details are given.

Primary and secondary amines can also interact in the palladium(0)-catalyzed dimerization of butadiene[134,138] (Chapter I, Section C,3,c).

$$2\text{CH}_2\text{=CHCH=CH}_2 \xrightarrow[(\text{Ph}_3\text{P})_2\text{Pd (maleic anhydride)}]{\text{RNH}_2} \text{RNH(CH}_2\text{CH=CH(CH}_2)_3\text{CH=CH}_2)_2$$

$$\xrightarrow[\text{R}_2\text{NH}]{(\text{Ph}_3\text{P})_2\text{PdC}_4\text{H}_2\text{O}_3} \text{R}_2\text{NH}\cdot\text{CH}_2\text{CH=CH(CH}_2)_3\text{CH=CH}_2$$

A general reaction to form allylamines,

$$\text{RCH=CHCH}_2\text{OR}' + \text{Et}_2\text{NH} \longrightarrow \text{RCH=CHCH}_2\text{NEt}_2 + \text{R}'\text{OH}$$

$$(\text{R} = \text{H, alkyl; R}' = \text{H, Ac, Ph})$$

is catalyzed by $Pd(acac)_2/PPh_3$[200,473] or by $(Ph_3P)_2Pd$(maleic anhydride).[201] Primary amines give diallylamines.[200]

A number of cases are known in which C–N bonds have been formed in a carbonylation reaction. Beck *et al.*[561] showed that a number of azides, including those of palladium(II) reacted with carbon monoxide to give isocyanates.

$$\text{L}_2\text{Pd(N}_3)_2 + 2\text{CO} \rightarrow \text{L}_2\text{Pd(NCO)}_2 + 2\text{N}_2$$

$$(\text{L} = \text{PPh}_3, \text{Bu}_3\text{P, piperidine})$$

Organic amines were also carbonylated to isocyanates.[93,562,563]

$$\text{RNH}_2 + \text{CO} + \text{PdCl}_2 \rightarrow \text{RNCO} + \text{Pd} + \text{HCl}$$

In the presence of allyl chloride, Tsuji and Iwamoto isolated allylpalladium chloride.[562] Stern and Spector[563] showed that the isocyanates did not arise via formamides, and also not via primary formation of phosgene. They suggested that a complex containing the amine, carbonyl, and chlorine, as well as palladium, was an intermediate.

Tsuji and Iwamoto[93] in another paper reported the isolation of ureas and oxamides from the above reaction at higher pressures of CO and higher temperatures.

$$RNH_2 + CO + PdCl_2 \rightarrow RNHCONHR + RNHCOCONHR$$

The true catalyst here was supposed to be the metal itself. Formamides were also isolated, particularly from lower aliphatic amines.

Haynes *et al.*[564] have reported the unusual carbonylation of secondary amines to formamides by carbon *dioxide* in the presence of metal catalysts. $(Ph_3P)_2PdCO_3$ was moderately effective but less so than cobalt or iridium complexes.

$$R_2NH + CO_2 + H_2 \rightarrow R_2NCOH + H_2O$$

Takahashi and Tsuji[41] were able to carbonylate the complex (**IV-7**) to the indazolinone.

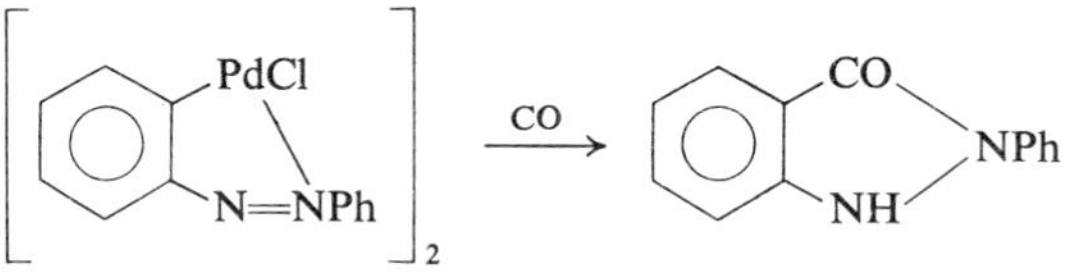

[Structure: dimeric ortho-palladated azobenzene complex [C6H4(PdCl)(N=NPh)]2 —CO→ indazolinone C6H4(CO)(NH)NPh]

(**IV-7**)

An ICI patent[565] has described the carbonylation of organic nitro compounds in alcohols to urethanes.

$$PhNO_2 + CO + MeOH \rightarrow PhNHCOOMe$$

Dichlorobis(triphenylphosphine)palladium appeared to be one of the best of a range of metal catalysts for this reaction.

Aromatic nitration can be catalyzed by palladium in one case at least.[436] Tissue and Downs reported that palladium black, when dissolved in acetic acid in the presence of nitrite, nitrated benzene to nitrobenzene. This may be mechanistically related to the chlorination of azobenzene (Section A,3).

An unusual reaction involving the formation of a tetrazole has been reported by Beck *et al.*[561]

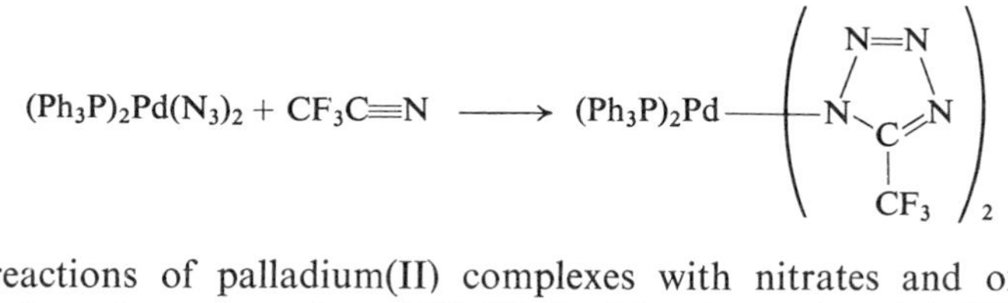

$$(Ph_3P)_2Pd(N_3)_2 + CF_3C{\equiv}N \longrightarrow (Ph_3P)_2Pd\left(N_4C{-}CF_3\right)_2$$

The reactions of palladium(II) complexes with nitrates and oxides of nitrogen have been investigated.[296,566] Braithwaite and Wright[566] isolated a complex, which they formulated as a cyanoacetate, from reaction of palladium(II) acetate and lithium nitrate in acetic acid.

$$Pd(OAc)_2 + LiNO_3 \xrightarrow{HOAc} NO_2 + Pd(CN)OAc \xrightarrow{py} (py)_2Pd(CN)_2 + (py)_2Pd(OAc)_2$$

This remarkable reaction obviously involves oxidation of acetate by nitrate; a mechanism involving the intermediate formation of methyl radicals, which reacted with NO, was suggested.

Isonitrilepalladium(II) complexes react with a variety of amines (and hydrazine) to give complexes containing carbene-type ligands[468] (Volume I, Chapter II, Section B,3) e.g.,

$$(PhNC)PdLCl_2 + p\text{-}MeC_6H_4NH_2 \longrightarrow (PhNH)(p\text{-}MeC_6H_4NH)C{-}PdLCl_2$$

Carbon–nitrogen bonds are broken during the oxidation of allylamines under the conditions of the Wacker reaction, e.g.,[251,334]

$$CH_2{=}CHCH_2NH_2 \xrightarrow{PdCl_2/H_2O} EtCHO + MeCOCHO$$

C. CARBON–SULFUR BONDS

Two reactions in which SO_2 is inserted have been reported. Klein[118] described the formation of ethyl *trans*-2-butenyl sulfone **(IV-13)** from ethylene and $PdCl_2$ in benzene containing traces of water. In water as a solvent, a Shell patent[119] claimed the formation of ethyl vinyl sulfone (Chapter I, Section C,1).

$$C_2H_4 + SO_2 \xrightarrow{PdCl_2/C_6H_6} EtSO_2CH_2CH{=}CHMe \quad \textbf{(IV-13)}$$

$$C_2H_4 + SO_2 \xrightarrow{PdCl_2/KCl/H_2O} EtSO_2CH{=}CH_2$$

O'Brien[567] showed that SO_2 reacted with bis(π-allyl)palladium to give a σ-bonded complex. The existence of two forms of this was deduced from the temperature variation of the NMR and IR spectra; **(IV-14a)** was present in greater amount at higher temperatures and **(IV-14b)** at lower temperatures.

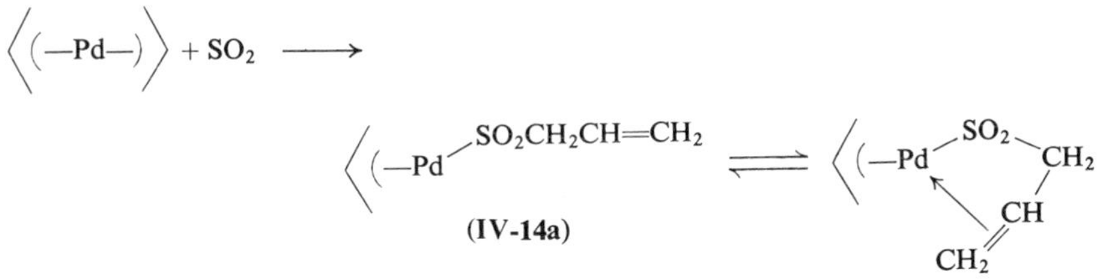

$$\left\langle\left(\text{—Pd—}\right)\right\rangle + SO_2 \longrightarrow$$

$$\left\langle\left(\text{—Pd}\right.\right.{-}SO_2CH_2CH{=}CH_2 \rightleftharpoons \left\langle\left(\text{—Pd}\right.\right.(SO_2CH_2CH{=}CH_2)$$

(IV-14a) (IV-14b)

Garves[46] has reported that arylsulfinates in the presence of palladium chloride lose SO_2 and give (presumably) arylpalladium complexes. These decompose, either on their own to biaryls, or react with a substrate (CO or olefin) to give the typical products arising from arylpalladium complexes.

$$ArSO_2Na + Na_2PdCl_4 \longrightarrow Ar{-}Ar$$

$$\xrightarrow{CO/MeOH} ArCOOMe + Ar_2CO$$

$$\xrightarrow{PhCH{=}CH_2} trans\text{-}PhCH{=}CHAr$$

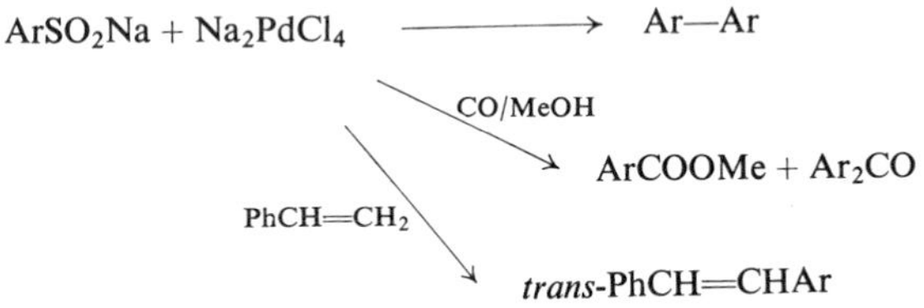

D. C–Si, Si–Si, AND H–Si

Takahashi *et al.*[140] have reported the codimerization of butadiene and trimethylsilane in the presence of a Pd(0) catalyst to give 1-trimethylsilyl-2,6-octadiene **(IV-15)**. This is in contrast to the normal codimerization of butadiene with Y–H (alcohols, amines, etc.) which gives the 1-substituted 2,7-octadiene (Chapter I, Section C,3,c).

$$Me_3SiH + 2CH_2{=}CHCH{=}CH_2 \xrightarrow{(Ph_3P)_2Pd\ (\text{maleic anhydride})} Me_3SiCH_2CH{=}CH(CH_2)_2CH{=}CHMe$$

(IV-15)

This palladium complex is also a catalyst for the normal hydrosilation of butadiene (without dimerization) by Cl_3SiH and Me_2PhSiH.

Some very curious reactions of di- and trisilanes with ethanol are also catalyzed by palladium chloride. They have been investigated by Yamamoto *et al.*[466]

$$Me_3SiSi(Me_2)Si(Me_2)CH{=}CH_2 + EtOH \xrightarrow{PdCl_2} Me_3SiOEt + Me_2Si(OEt)_2 +$$

$$CH_2{=}CH(Me_2)SiOEt + EtMe_2SiOEt + Me_3SiSiMe_2OEt + Me_3SiSi(Me_2)SiMe_2OEt$$

Sommer and Lyons[568] have shown that optically active silanes (R_3Si^*H) undergo solvolysis by X–H (alcohols, amines) to give $R_3{}^*SiX$ with complete inversion of configuration on a palladium metal catalyst. Other metals are less efficient.

Chapter V

Heterogeneous Reactions

A detailed discussion of reactions on palladium metal is quite outside the scope of this book, but it is appropriate to consider briefly and comparatively the reactions which occur homogeneously with those for which the metal is active.

The transition metals, particularly those of group VIII, are able to induce a large number of reactions and many studies of such systems have been reported. They have been discussed in considerable detail in the excellent book by Bond.[569]

There are also heterogeneously catalyzed reactions which do not, apparently, involve the metal. For example, a number of patents reports the gas-phase syntheses of acetaldehyde,[570–574] vinyl acetate,[271,414,575] vinyl ether,[414] and acetals[414] on a solid catalyst consisting of a palladium(II) salt together with an oxidizer ($CuCl_2$, V_2O_5, etc.) impregnated on a support such as alumina, carbon, silica gel, or a zeolite. It seems likely that such reactions involve the same processes which occur during the analogous homogeneous reactions in solution. In the vinyl acetate synthesis this type of heterogeneous catalysis offers some advantages over the homogeneous reaction (Chapter II, Section B,1).

There is, in fact, no great similarity between the reactions which have been established for palladium *metal* and those which occur homogeneously either with Pd(II) or Pd(0), although in the latter case the published work to date has only been fragmentary.

For example, palladium, in common with most other transition metals, is effective in catalyzing the hydrogenation of unsaturated organic compounds. Very few reactions where soluble Pd complexes catalyze such reactions have been reported. On the other hand, double bond isomerizations of various types are readily catalyzed both by palladium surfaces and by complexes in solution, though under widely different conditions.

Oxidation also occurs on palladium metal, but the products are often different; for example, ethylene is usually oxidized right through to carbon dioxide and water.[576] Formation of acetaldehyde or acetic acid, except in the presence of gold, only takes place to a limited extent.[283,384,577–579]

A brief account of the more important reactions catalyzed by palladium metal follows.

A. HYDROGENATION, HYDROGENOLYSIS, ISOMERIZATION, AND HYDROGEN-TRANSFER REACTIONS

Palladium is a reasonably good catalyst for the hydrogenation of multiple bonds, and is useful because of its selectivity. Thus it has low reactivity toward aromatic rings by comparison with some other metals. It is also very effective in selectively hydrogenating acetylenes to *cis*-olefins; indeed hydrogenation of the resultant olefin is completely repressed by poisoning the catalyst with lead acetate and quinoline (Lindlar's catalyst) or with dimethyl sulfide. Diolefins behave similarly.[569,580]

Aromatic nitro compounds are reduced to amines, and ketones to alcohols and the saturated hydrocarbon, but in this case other metals are much more effective.

Palladium is also good at causing hydrogenolysis, particularly of carbon–halogen bonds. An approximate order of reactivity is[581] $CH_2{=}CHCl > C_2H_5Cl > CH_3Cl > n\text{-}C_3H_7F > i\text{-}C_3H_7F$. A specific case of this reaction is the well-known Rosenmund reduction of acyl halides to aldehydes, which occurs on palladium black deposited on barium sulfate.

$$RCOCl + H_2 \rightarrow RCHO + HCl$$

Hydrogenolysis of benzylic alcohols has also been reported.[531c]

Palladium is also able to catalyze double-bond isomerization, although a higher temperature than for the homogeneously catalyzed process is usually needed.[426, 508]

Hydrogenation, hydrogenolysis, isomerization, hydrogen-transfer, and dehydrogenation reactions all occur simultaneously and have common mechanistic paths (see below). Palladium is accordingly useful in catalyzing dehydrogenation and hydrogen-transfer reactions. For example, aromatic nitro compounds are reduced to amines by palladium in cyclohexane in the absence of molecular hydrogen. Similarly, it catalyzes the disproportionation of cyclohexene to benzene and cyclohexane.[516]

A property not observed at all as yet in homogeneous systems is the reaction with saturated hydrocarbons. A number of reactions occur on metal surfaces and, of these, the exchange with deuterium on palladium has been widely studied.[582–584]

Formic acid is also decomposed in the gas phase on palladium surfaces; in this case, however, the products are hydrogen and carbon dioxide[569] (compare Chapter II, Section G). The decomposition of formate esters on Pd/C has been reported recently by Matthews *et al.*[585]

B. CARBONYLATION AND DECARBONYLATION

Palladium, usually supported on carbon, has been used as a catalyst for the carbonylation of a large number of olefins and acetylenes by Tsuji and his co-workers.[72, 89–91, 252] Even quite simple reactants, however, give very complex mixtures; some of these reactions are discussed in Chapter I, Section B. Tsuji and Iwamoto[93] have also reported the carbonylation of primary amines to oxamides and formamides.

$$RNH_2 + CO \xrightarrow{Pd/C} RNHCOCONHR + RNHCOH$$

The palladium-catalyzed decarbonylation of aldehydes[254, 255] and acyl halides[252, 253] has been reported. These reactions occurred in the liquid phase, and Tsuji *et al.*[252] observed that metallic palladium was soluble to some extent in the acyl chloride.

Carbonylation does not appear to have been observed in the gas phase, but the decarbonylation of aldehydes to hydrocarbons is known.[569]

A mechanism for the liquid-phase reactions was proposed[252] and is presented in Chapter I, Section F,3.

C. OXIDATION

Ethylene reacts with oxygen on a number of metal films to give carbon dioxide and water. Traces of acetic acid and anhydride are also observed; these are formed by a different route and are not intermediates.[576] However, Gerberich *et al.*[283,578] found that a gold–palladium alloy increased the selectivity to partial oxidation considerably; acetaldehyde then became a major product. Blake *et al.*[577] have also found that acetaldehyde was formed on anodic oxidation of ethylene in water using a Pd–Au electrode.

A palladium–gold on silica catalyst has been described in two patents[384,579] as active for the formation of vinyl acetate from ethylene in the gas phase. However, other catalysts, apparently not containing gold, have also been proposed for the same process.[101,271,289,369,382,398,399,586,587]

The gas-phase reaction of oxygen with hydrogen and carbon monoxide is catalyzed by palladium; the former is retarded by addition of gold, whereas the latter is accelerated.[569] The changes in catalytic activity on alloying with gold are not understood, but it has been suggested that the gold acts by filling the palladium d band with s electrons.

The decomposition of hydrogen peroxide also shows highest activity for a Pd–Au alloy.[588]

D. CARBON–CARBON BOND FORMATION

Palladium is active as a promoter in the catalytic cracking of high molecular weight hydrocarbons, but platinum is, of course, superior.[569] In gas-phase reactions chemisorbed ethylene and acetylene oligomerize in the presence of hydrogen.[569] Bryce-Smith[195] has reported the cyclotrimerization of dimethylacetylenedicarboxylate and a number of monosubstituted acetylenes. Linear polymers were also produced from the latter reactions. These oligomerization reactions proceeded readily in benzene in the presence of palladium on charcoal; they should be compared with the homogeneously catalyzed reactions (Chapter I, Section C,3,d).

Other coupling reactions reported include the formation of dicyclohexyl and cyclohexylbenzene using a catalyst containg palladium, tungsten, and fluoride on silica–alumina.[589] An acidic substrate is necessary for this reaction and a nickel-based catalyst is somewhat more active.

$$C_6H_6 + H_2 \longrightarrow \text{(phenylcyclohexane)} + \text{(bicyclohexyl)}$$

Fujiwara *et al.*[234] have reported the vinylation of aromatics using palladium metal. This reaction is obviously similar to those carried out homogeneously (Chapter I, Section D).

E. MISCELLANEOUS

Other reactions promoted by palladium metal include the hydrocyanation of olefins,[35,590]

$$C_2H_4 + HCN \rightarrow CH_3CH_2CN$$

and the solvolysis of optically active silanes (RR′R″SiH) which occurs with complete inversion.[568]

A patent has reported the formation of vinyl chloride and ethyl chloride in small amounts from the oxidation of ethylene in the presence of HCl on a palladium catalyst.[551]

Palladium chloride also catalyzes the reaction of phosgene with ethylene oxide.[470]

$$\text{(ethylene oxide)} + COCl_2 \xrightarrow{PdCl_2/Al_2O_3} ClCH_2CH_2OC(=O)Cl + (ClCH_2CH_2O)_2CO$$

F. MECHANISTIC CONSIDERATIONS

Heterogeneous reactions in the liquid phase have been little studied, and although reactions at metal surfaces in the gas phase have been the subject of numerous investigations, our understanding even of them is very limited. It is, for example, not even clear whether reactions occur uniformly over the whole surface or only at "active sites", corresponding to faults, dislocations, and similar features.

Infrared spectroscopy has been of help in identifying species adsorbed on surfaces, but here again it is not clear just what the role of these species is in the reactions which occur.

Little[591] has summarized the evidence of a number of workers who have shown that molecules such as CO, H_2, and N_2 give rise to characterisitic absorption bands in the infrared which are very close to those reported for molecular complexes containing carbonyl, hydride, or nitrogen ligands. It therefore seems probable that these entities when chemisorbed on a metal surface are electronically similar to and will behave analogously to such ligands in complexes. For example, carbon monoxide on supported palladium gives rise to bands at 2050(w), 1920(s), and 1827(sh) cm^{-1}. The first is assigned to a terminal Pd–CO, the latter two to bridging Pd–C(=O)–Pd groups. Nickel gave very similar spectra, but platinum gave only a strong band at 2070 cm^{-1} (terminal Pt–CO).

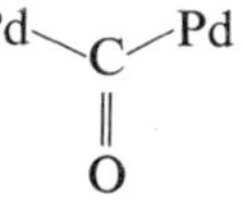

The situation for hydrocarbons is more complex since reactions other than simple adsorption occur.[569] Most interpretations of the infrared data rely on the C–H stretching region, and are therefore subject to some disagreement. The spectrum of ethylene chemisorbed on palladium supported on silica showed bands at 2870, 2970, and 3030 cm^{-1}. The last band is interpreted in terms of a metal-bonded species such as

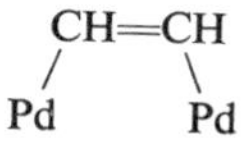

$$\begin{array}{ccc} & CH{=}CH & \\ / & & \backslash \\ Pd & & Pd \end{array}$$

whereas the others may arise from saturated groups, for example, Pd–CH_2CH_3. The olefinic band disappeared and lower frequency bands (at 2850, 2920, and 2950 cm^{-1}) increased dramatically on introducing hydrogen to the system. Similar results were obtained with nickel. Since the species to which these bands are due did not desorb readily from the surface, they are thought *not* to be the intermediates involved in the hydrogenation reaction to give ethane.[591] Ethane was adsorbed on a nickel surface to give a species with a spectrum identical to that of chemisorbed ethylene.

The situation for acetylene adsorbed on palladium is similarly complex and the structures

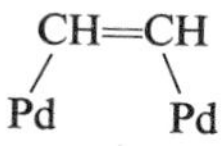

$$\begin{array}{ccc} & CH{=}CH & \\ / & & \backslash \\ Pd & & Pd \end{array} \quad \text{and} \quad \begin{array}{ccc} H & & H \\ & \diagdown C \diagup & \\ & \| & \\ & C & \\ Pd \diagup & & \diagdown Pd \end{array}$$

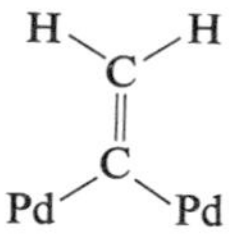

have both been proposed. After addition of hydrogen the infrared spectrum was in agreement with the structure

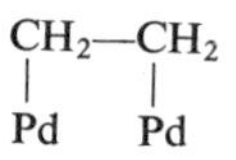

$$\begin{array}{cc} CH_2{-}{-}CH_2 \\ | \quad\quad | \\ Pd \quad\; Pd \end{array}$$

Evidence in support of the dissociative adsorption of ethylene is that ethane is observed; this implies an overall reaction such as[569]

$$C_2H_4 + \text{metal} \rightarrow C_2H_6 + C_2H_2\text{metal}$$

The reaction of ethylene with deuterium on a palladium surface has been discussed by Bond, who suggested that processes such as the following occur,[569, 592]

$$CH_2{=}CH_2 + * \rightleftharpoons CH_2\underset{\overset{|}{*}}{=}CH_2$$

$$D_2 + 2* \rightleftharpoons 2\underset{\overset{|}{*}}{D}$$

$$CH_2\underset{\overset{|}{*}}{=}CH_2 + \underset{\overset{|}{*}}{D} \rightleftharpoons \underset{\overset{|}{*}}{CH_2}{-}CH_2D + *$$

$$CH_2\underset{\overset{|}{*}}{=}CH_2 + D_2 + * \rightleftharpoons \underset{\overset{|}{*}}{CH_2}{-}CH_2D + \underset{\overset{|}{*}}{D}$$

$$CH_2\underset{\overset{|}{*}}{=}CH_2 + \underset{\overset{|}{*}}{CH_2}{-}CH_2D \rightarrow \underset{\overset{|}{*}}{CH_2}CH_3 + CH_2\underset{\overset{|}{*}}{=}CHD$$

$$\underset{\overset{|}{*}}{CH_2}{-}CH_2D + \underset{\overset{|}{*}}{D} \rightarrow DCH_2CH_2D + 2*$$

$$\underset{\overset{|}{*}}{CH_2}{-}CH_2D + D_2 \rightarrow DCH_2{-}CH_2D + \underset{\overset{|}{*}}{D}$$

$$2\underset{\overset{|}{*}}{CH_2}{-}CH_2D \rightarrow CH_3CH_2D + CH_2\underset{\overset{|}{*}}{=}CHD + *$$

where $\underset{\overset{|}{*}}{X}$ represents a group X attached to an atom on the surface of the metal.

Since $CH_2{=}CHD$ and ethane can react further with the metal, all possible isotopic species will be found. In fact, in one experiment Bond *et al.*[592] found the following amounts of deuteroethanes and -ethylenes after 5% conversion had occurred: C_2H_3D (40.8); $C_2H_2D_2$ (7.9); C_2HD_3 (2.2); C_2D_4 (0.8); C_2H_6 (21.6); C_2H_5D (17.6); $C_2H_4D_2$ (7.3); $C_2H_3D_3$ (1.2); and $C_2H_2D_4$ (0.5%).

Bond *et al.* now consider that the olefins are π-bonded to the metal in the manner usual for metal complexes (Volume I, Chapter III, Section D). Rooney[583] and Siegel[593] have interpreted their results similarly and have also invoked the intermediacy of π-allylic and similar complexes in their reactions. However, there is still some evidence and a considerable weight of opinion (for example, see Burwell[582]) in favor of an olefin needing two adjacent metal atoms to attach itself to. This situation is well known for π-acetylene complexes (Volume I, Chapter III, Section A), but olefins have only one set of π and π^* orbitals and can therefore only bond to one metal atom in this way. Therefore, if an olefin is attached to two metal atoms, as has been proposed,[569] this implies a di-σ-bonded structure. The question of how many surface metal atoms are needed to bind an olefin is important and still appears unresolved.

A similar problem arises that, in considering reactions such as hydrogenation at gas–metal interfaces, all workers are agreed in requiring coordination of the organic moiety and the hydrogen (or deuterium) atom to need two separate, but presumably adjacent, atoms. This contrasts with the experience of workers in the field of hydrogenation under homogeneous conditions in solution who find that the experimental evidence is best accounted for in terms of both the hydrogen (as metal hydride) and the olefin, or other reactive substrate, being coordinated to the *same* metal atom.

Mechanisms such as those suggested by Bond, above, imply that the processes which occur on a metal surface are similar to those which occur under homogeneous conditions. If this is indeed so, it leads to the intriguing possibility that the dissociative chemisorption of saturated hydrocarbons on metal surfaces will also eventually be mimicked by homogeneous catalysts.

The very high reactivity of clean films of palladium and neighboring metals in these and other reactions quite belies the so-called "nobility" of these metals. In fact, it suggests the metals are highly reactive toward oxidation reactions of all types, but that this is impeded in some way by a deactivation of the surface under "normal" conditions. In the absence of a vacant coordination site no reaction can occur; presumably in the liquid phase, even with a "clean" surface competition of solvent for the vacant positions minimizes reaction. However, a number of reactions in which metallic palladium is solubilized are known.[594–596]

Glossary

Ac	Acetyl	$CH_3CO—$
acac	Acetylacetonate	
bipy	Bipyridyl, 2,2′-dipyridyl	
B.M.	Bohr magneton (magnetic moment)	
Bu	Butyl	$C_4H_9—$
Cp	Cyclopentadienyl	
COD	Cyclooctadiene	C_8H_{12}
COT	Cyclooctatetraene	
D	Debye (dipole moment)	
diars	*o*-Phenylenebis(dimethylarsine)	$AsMe_2$ / $AsMe_2$
dien	Diethylenetriamine	$NH_2CH_2CH_2NHCH_2CH_2NH_2$
diphos	1,2-Bis(diphenylphosphino)ethane	$Ph_2PCH_2CH_2PPh_2$
DMAC	*N,N′*-Dimethylacetamide	Me_2NCOMe
DMF	*N,N′*-Dimethylformamide	Me_2NCHO
DMSO	Dimethyl sulfoxide	Me_2SO
DTA	Differential thermal analysis	
E	Usually phosphorus, arsenic, or antimony	
e.s.d.	Estimated standard deviation †	
ESR	Electron spin resonance (or electron paramagnetic resonance)	

† E.s.d's are expressed in parentheses after the bond length. A Pd–C bond length of 2.132(7)Å (or 2.132 ± 0.007Å) implies that the true value is about 70% certain to be within this range (2.125–2.139Å) and 99% certain to be within *three* times the e.s.d. (2.111–2.153Å)

Et	Ethyl	CH_3CH_2—
Et_4dien	*N,N,N,N*-Tetraethyldiethylenetriamine	$Et_2NCH_2CH_2NHCH_2CH_2NEt_2$
Me	Methyl	CH_3—
NMR	Nuclear magnetic resonance, usually of protons, ^1H	
NQR	Nuclear quadrupole resonance	
Ph	Phenyl	
o-phen	*o*-Phenanthroline	N N
PMR	Proton magnetic resonance	
Pr	Propyl	C_3H_7—
py	Pyridine	N
R	Organic radical	
R_f	Perfluoroalkyl or -aryl	
TAS	2,6,10-Trimethyl-2,6,10-triarsaundecane	$Me_2As(CH_2)_3As(Me)(CH_2)_3AsMe_2$
X	Usually halide	

Bibliography[†]

1. J. Tsuji and H. Takahashi, *J. Amer. Chem. Soc.* **87**, 3275 (1965).
2. J. Tsuji and H. Takahashi (Toyo Rayon Co. Ltd.), Jap. Pat. 23,180/67; *Chem. Abstr.* **69**, 52305x (1968).
3. H. Takahashi and J. Tsuji, *J. Amer. Chem. Soc.* **90**, 2387 (1968).
4. B. F. G. Johnson, J. Lewis, and M. S. Subramanian, *J. Chem. Soc., A* p. 1993 (1968).
5. D. R. Coulson, *J. Amer. Chem. Soc.* **91**, 201 (1969).
6. C. B. Anderson and B. J. Burreson, *J. Organometal. Chem.* **7**, 181 (1967); *Chem. Ind. (London)* p. 620 (1967).
7. G. D. Shier, *Amer. Chem. Soc., Div. Petrol. Chem., Prepr.* **14**, B123 (1969).
8. R. F. Heck, *J. Amer. Chem. Soc.* **90**, 317 (1968).
9. R. F. Heck, *J. Amer. Chem. Soc.* **90**, 5542 (1968).
10. D. Medema and R. van Helden, *Amer. Chem. Soc., Div. Petrol. Chem., Prepr.* **14**, B92 (1969).
11. D. Medema, R. van Helden, and C. F. Kohll, *Abstr., Inorg. Chim. Acta Meet., 1968* p. E3 (1968); *Inorg. Chim. Acta* **3**, 255 (1969).
12. A. Bright, B. L. Shaw, and G. Shaw, *Amer. Chem. Soc., Div. Petrol. Chem., Prepr.* **14**, B81 (1969).
13. Y. Takahashi, S. Sakai, and Y. Ishii, *Chem. Commun.* p. 1092 (1967); Y. Takahashi, K. Tsukiyama, S. Sakai, and Y. Ishii, *Tetrahedron Lett.* 1913 (1970).
14. Y. Takahashi, S. Sakai, and Y. Ishii, *J. Organometal. Chem.* **16**, 177 (1969).
15. J. Tsuji, H. Takahashi, and M. Morikawa, *Tetrahedron Lett.* p. 4387 (1965).
16. J. Tsuji, H. Takahashi, and M. Morikawa, *Kogyo Kagaku Zasshi* **69**, 920 (1966).
17. H. Okada and H. Hashimoto, *Kogyo Kagaku Zasshi* **70**, 2152 (1967).
18. T. Saegusa, T. Tsuda, and K. Nishijima, *Tetrahedron Lett.* p. 4255 (1967).
19. R. F. Heck, *J. Amer. Chem. Soc.* **90**, 5518 (1968).
20. R. F. Heck, *J. Amer. Chem. Soc.* **90**, 5538 (1968).
21. R. F. Heck, personal communication (1970).
22. R. F. Heck, *J. Amer. Chem. Soc.* **90**, 5526 (1968).
23. R. F. Heck, *J. Amer. Chem. Soc.* **90**, 5531 (1968).
24. R. F. Heck, *J. Amer. Chem. Soc.* **90**, 5535 (1968).
25. R. F. Heck, *J. Amer. Chem. Soc.* **91**, 6707 (1969).
26. P. Arthur, D. C. England, B. C. Pratt, and G. M. Whitman, *J. Amer. Chem. Soc.* **76**, 5364 (1954).
27. W. C. Drinkard, Belg. Pat., 698,322 (1967); W. C. Drinkard and R. V. Lindsey, Belg. Pat., 698,333 (1967).
28. E. S. Brown and E. A. Rick, *Chem. Commun.* p. 112 (1969).

† A number of references are given to Russian journals in their English versions. Russ. Chem. Rev. is the translation of Uspekhi Khimii (Chemical Society, London); Russ. J. Inorg. Chem. is the translation of Zh. Neorg. Khim. (Chemical Society, London); Dokl. Akad. Nauk SSSR (English translation) refers to Proceedings of the Academy of Sciences of the USSR, translated by Consultants Bureau, New York; Izv. Akad. Nauk SSSR (English translation) refers to Bulletin of the Academy of Sciences of the USSR, Division of Chemical Sciences, translated by Consultants Bureau. J. Struct. Chem. USSR is the translation of Zh. Strukt. Khim. (Consultants Bureau), and "Kinetics and Catalysis" of Kinetika i Kataliz (Consultants Bureau). In all these cases the page referred to is that of the English translation.

29. E. S. Brown and E. A. Rick, *Amer. Chem. Soc., Div. Petrol. Chem., Prepr.* **14**, B29 (1969).
30. H. Singer and G. Wilkinson, *J. Chem. Soc., A* p. 2516 (1968).
31. W. C. Drinkard, D. R. Eaton, J. P. Jesson, and R. V. Lindsey, *Inorg. Chem.* **9**, 392 (1970); C. A. Tolman, *J. Amer. Chem. Soc.* **92**, 4217 (1970).
32. R. A. Schunn, *Inorg. Chem.* **9**, 394 (1970).
33. Y. Odaira, T. Oishi, T. Yukawa, and S. Tsutsumi, *J. Amer. Chem. Soc.* **88**, 4105 (1966).
34. Farbwerke Hoechst, Neth. Pat., 6,505,608 (1965); *Chem. Abstr.* **64**, 12557e (1966).
35. Asahi Chemical Industry Co., Belg. Pat., 670,276 (1966); *Chem. Abstr.* **65**, 15241a (1966).
36. R. F. Heck, *in* "Organic Syntheses via Metal Carbonyls" (I. Wender and P. Pino, eds.), p. 373. Wiley (Interscience), New York, 1968.
37. K. Noack and F. Calderazzo, *J. Organometal. Chem.* **10**, 101 (1967); K. Noack, M. Ruch, and F. Calderazzo, *Inorg. Chem.* **7**, 345 (1968).
38. P. Cossee, *Rec. Trav. Chim. Pays-Bas* **85**, 1151 (1966).
39. G. Booth and J. Chatt, *Proc. Chem. Soc., London* p. 67 (1961).
40. G. Booth and J. Chatt, *J. Chem. Soc., A* p. 634 (1966).
41. H. Takahashi and J. Tsuji, *J. Organometal. Chem.* **10**, 511 (1967).
42. J. K. Stille and L. F. Haines, *J. Amer. Chem. Soc.* **92**, 1798 (1970).
43. S. D. Robinson and B. L. Shaw, *J. Chem. Soc., London* p. 5002 (1964).
44. J. Tsuji and S. Hosaka, *J. Polym. Sci., Part B* **3**, 703 (1965).
45. P. M. Henry, *Tetrahedron Lett.* p. 2285 (1968).
46. K. Garves, *J. Org. Chem.* **35**, 3273 (1970).
47. National Distillers and Chem. Corp., Brit. Pat., 1,149,359 (1969); *Platinum Metals Rev.* **13**, 123 (1969).
48. J. Tsuji, J. Kiji, and M. Morikawa, *Tetrahedron Lett.* p. 1811 (1963).
49. J. Tsuji, J. Kiji, and S. Hosaka, *Tetrahedron Lett.* p. 605 (1964).
50. J. Tsuji, J. Kiji, S. Imamura, and M. Morikawa, *J. Amer. Chem. Soc.* **86**, 4350 (1964).
51. J. Tsuji, S. Imamura, and J. Kiji, *J. Amer. Chem. Soc.* **86**, 4491 (1964).
52. J. Tsuji and T. Susuki, *Tetrahedron Lett.* p. 3027 (1965).
53. J. Tsuji and S. Imamura, *Bull. Chem. Soc. Jap.* **40**, 197 (1967).
54. R. Long and G. H. Whitfield, *J. Chem. Soc., London* p. 1852 (1964).
55. G. Chiusoli, *Chim. Ind.* (*Milan*) **41**, 503 (1959).
56. W. T. Dent, R. Long, and G. H. Whitfield, *J. Chem. Soc., London* p. 1588 (1964).
57. Shell Internationale Research Maatschappij N.V., Brit. Pat., 1,080,867 (1965); *Chem. Abstr.* **63**, 499c (1965).
58. R. Long and G. H. Whitfield (ICI Ltd.), Brit. Pat., 987,274 (1965); *Chem. Abstr.* **62**, 16065h (1965).
59. I. L. Mador and J. A. Scheber (National Distillers and Chem. Corp.), Fr. Pat., 1,419,758 (1965); *Chem. Abstr.* **65**, 13553d (1966).
60. H. G. Volger, K. Vrieze, J. W. F. M. Lemmers, A. P. Praat, and P. W. N. M. van Leeuwen, *Inorg. Chim. Acta* **4**, 435 (1970).
61. J. Tsuji, M. Morikawa, and J. Kiji, *J. Amer. Chem. Soc.* **86**, 4851 (1964).
62. J. Tsuji, M. Morikawa, and J. Kiji, *Tetrahedron Lett.* p. 1061 (1963).
63. A. U. Blackham (National Distillers and Chem. Corp.), U.S. Pat., 3,119,861 (1964); *Chem. Abstr.* **60**, 9155h (1964).
64. N. V. Kutepow, K. Bittler, D. Neubauer, and H. Reiss, Ger. Pat., 1,237,116 (1967); *Chem. Abstr.* **68**, 21550y (1968).
65. K. Bittler, N. V. Kutepow, D. Neubauer, and H. Reiss, *Angew. Chem., Int. Ed. Engl.* **7**, 329 (1968).

66. Badische Aniline u. Soda-Fabrik AG, Neth. Pat., 6,409,121 (1965); *Chem. Abstr.* **63**, 14726b (1965).
67. Badische Aniline u. Soda-Fabrik AG, Neth. Pat., 6,516,439 (1966); *Chem. Abstr.* **65**, 15249a (1966).
68. N. V. Kutepow, D. Neubauer, and K. Bittler (BASF), Ger. Pat., 1,229,089 (1966); *Chem. Abstr.* **66**, 37510e (1967).
69. J. Tsuji and S. Hosaka (Toyo Rayon Co. Ltd.), Jap. Pat., 5699 (1965); *Chem. Abstr.* **62**, 16085c (1965).
70. J. Tsuji and T. Nogi, *Bull. Chem. Soc. Jap.* **39**, 146 (1966).
71. J. Tsuji, S. Hosaka, T. Kiji, and T. Susuki, *Bull. Chem. Soc. Jap.* **39**, 141 (1966).
72. J. Tsuji, M. Morikawa, and J. Kiji, *Tetrahedron Lett.* p. 1437 (1963).
73. ICI Ltd., Neth. Pat., 6,511,995 (1966); *Chem. Abstr.* **65**, 7064f (1966).
74. J. Tsuji, N. Iwamoto, and M. Morikawa, *Bull. Chem. Soc. Jap.* **38**, 2213 (1965).
75. D. M. Fenton (Union Oil Co.), *Chem. Eng. News* Sept. 22, 1969, p. 72.
76. S. Brewis and P. R. Hughes, *Chem. Commun.* p. 489 (1965).
77. S. Brewis and P. R. Hughes, *Chem. Commun.* p. 6 (1966).
78. S. Brewis and P. R. Hughes, *Amer. Chem. Soc., Div. Petrol. Chem., Prepr.* **14**, B170 (1969).
79. S. Brewis and P. R. Hughes, *Chem. Commun.* p. 71 (1967).
80. S. Brewis and P. R. Hughes, *Chem. Commun.* p. 157 (1965).
81. J. Tsuji and S. Hosaka, *J. Amer. Chem. Soc.* **87**, 4075 (1965).
82. C. Bordenca and W. E. Marsico, *Tetrahedron Lett.* p. 1541 (1967).
83. T. Susuki and J. Tsuji, *Bull. Chem. Soc. Jap.* **41**, 1954 (1968).
84. J. Tsuji, M. Morikawa, and N. Iwamoto, *J. Amer. Chem. Soc.* **86**, 2095 (1964).
85. J. Tsuji, M. Morikawa, and N. Konochi (Toyo Rayon Co. Ltd.), Jap. Pat., 24,608 (1965); *Chem. Abstr.* **64**, 4946e (1966).
86. G. Jacobsen and H. Spaethe (Farbwerke Hoechst AG), Ger. Pat., 1,138,760 (1962); *Chem. Abstr.* **58**, 6699f (1963).
87. J. Tsuji and T. Nogi, *J. Amer. Chem. Soc.* **88**, 1289 (1966).
88. H. W. Sternberg, J. G. Shukys, C. D. Donne, R. Markby, R. A. Friedel, and I. Wender, *J. Amer. Chem. Soc.* **81**, 2399 (1959); O. S. Mills and G. Robinson, *Proc. Chem. Soc., London* p. 156 (1959).
89. J. Tsuji and T. Nogi, *Tetrahedron Lett.* p. 1801 (1966).
90. J. Tsuji and T. Nogi, *J. Org. Chem.* **31**, 2641 (1966).
91. T. Nogi and J. Tsuji, *Tetrahedron* **25**, 4099 (1969).
92. J. Tsuji, M. Morikawa, and J. Kiji, *Tetrahedron Lett.* p. 817 (1965).
93. J. Tsuji and N. Iwamoto, *Chem. Commun.* p. 380 (1966).
94. L. M. Vallarino and G. Santarella, *Gazz. Chim. Ital.* **94**, 252 (1964).
95. P. M. Maitlis and M. L. Games, *J. Amer. Chem. Soc.* **85**, 1887 (1963).
96. D. M. Fenton and P. J. Steinwand (Union Oil Co.), U.S. Pat., 3,393,136 (1968); *Platinum Metals Rev.* **13**, 43 (1969).
97. S. Usami, T. Kondo, K. Nishimura, and Y. Koga, *Bull. Chem. Soc. Jap.* **42**, 2961. (1969); S. Usami, K. Nishimura, T. Koyama, and S. Fukushi, *ibid.* p. 2966.
98. J. Smidt, W. Hafner, R. Jira, J. Sedlmeier, R. Sieber, R. Rüttinger, and H. Kojer, *Angew. Chem.* **71**, 176 (1959).
99. J. Smidt, W. Hafner, and J. Sedlmeier (Consortium für Elektrochemische Industrie GmbH.), Brit. Pat., 887,362 (1962); *Chem. Abstr.* **58**, 3521e (1963).
100. A. U. Blackham (National Distillers and Chemical Corp.), U.S. Pat., 3,194,800 (1965); *Chem. Abstr.* **63**, 8518f (1965).
101. Union Oil of California, Brit. Pat., 1,032,325 (1966); *Chem. Abstr.* **65**, 5372h (1966).

102. W. D. Schaeffer (Union Oil Co. of California), U.S. Pat., 3,285,970 (1966); *Chem. Abstr.* **66**, 37437m (1967).
103. H. S. Klein (Shell Oil Co.), U.S. Pat., 3,354,236 (1967); *Chem. Abstr.* **68**, 21491e (1968).
104. J. T. van Gemert and P. R. Wilkinson, *J. Phys. Chem.* **68**, 645 (1964).
105. Y. Kusunoki, R. Katsuno, N. Hasegawa, S. Kurematsu, Y. Nagao, K. Ishii, and S. Tsutsumi, *Bull. Chem. Soc. Jap.* **39**, 2021 (1966).
106. A. D. Ketley, L. P. Fisher, A. J. Berlin, C. R. Morgan, E. H. Gorman, and T. R. Steadman, *Inorg. Chem.* **6**, 657 (1967).
107. F. Conti, G. Pregaglia, and R. Ugo, *Amer. Chem. Soc., Div. Petrol. Chem., Prepr.* **14**, B39 (1969).
108. K. Kawamoto, T. Imanaka, and S. Teranishi, *Bull. Chem. Soc. Jap.* **42**, 2688 (1969).
109. A. Aguilo and L. Stautzenberger, *Chem. Commun.* p. 406 (1969).
110. J. C. Crano, E. K. Fleming, and G. M. Trenta, *J. Amer. Chem. Soc.* **90**, 5036 (1968).
111. M. G. Barlow, M. J. Bryant, R. N. Haszeldine, and A. G. Mackie, *J. Organometal. Chem.* **21**, 215 (1970).
112. K. Kawamoto, T. Imanaka, and S. Teranishi, *Kogyo Kagaku Zasshi* **72**, 1612 (1969); *Bull. Chem. Soc. Jap.* **43**, 2512 (1970).
113. U.S. Dept. of the Interior, Bureau of Mines, *Miner. Yearb.* Vol. I/II, p. 341 (1966).
114. E. M. Wise, "Palladium: Recovery, Properties and Applications." Academic Press, New York, 1968.
115. F. J. Weigert, R. L. Baird, and J. R. Shapley, *J. Amer. Chem. Soc.* **92**, 6630 (1970).
116. K. Dunne and F. J. McQuillin, *J. Chem. Soc., C* p. 2203 (1970).
117. A. D. Ketley and J. A. Braatz, *J. Polym. Sci., Part B* **6**, 341 (1968).
118. H. S. Klein, *Chem. Commun.* p. 377 (1968).
119. Shell Oil Co., U.S. Pat., 3,448,159 (1969); *Platinum Metals Rev.* **14**, 39 (1970).
120. W. Hafner, H. Prigge, and J. Smidt, *Justus Liebigs Ann. Chem.* **693**, 109 (1966).
121. W. H. Urry and M. B. Sullivan, *Amer. Chem. Soc., Div. Petrol. Chem., Prepr.* **14**, B131 (1969).
122. T. Hosokawa, I. Moritani, and S. Nishioka, *Tetrahedron Lett.* p. 3833 (1969).
123. P. Mushak and M. A. Battiste, *Chem. Commun.* p. 1146 (1969).
124. W. Schneider, *Amer. Chem. Soc., Div. Petrol. Chem., Prepr.* **14**, B89 (1969).
125. J. M. Rowe and D. A. White, *J. Chem. Soc., A* p. 1451 (1967).
126. R. Cramer, *J. Amer. Chem. Soc.*, **89**, 1633 (1967).
127. P. M. Maitlis and F. G. A. Stone, *Proc. Chem. Soc., London* p. 330 (1962).
128. P. M. Maitlis, *Advan. Organometal. Chem.* **4**, 95 (1966).
129. V. R. Sandel and H. H. Freedman, *J. Amer. Chem. Soc.* **90**, 2059 (1968).
130. R. C. Cookson and D. W. Jones, *Proc. Chem. Soc., London* p. 115 (1963).
131. R. C. Cookson and D. W. Jones, *J. Chem. Soc.* p. 1881 (1965).
132. R. P. Hughes and J. Powell, *J. Organometal. Chem.* **20**, P.17 (1969).
133. G. D. Shier, *J. Organometal. Chem.* **10**, P15 (1967).
134. S. Takahashi, T. Shibano, and N. Hagihara, *Bull. Chem. Soc. Jap.* **41**, 454 (1968).
135. S. Takahashi, T. Shibano, and N. Hagihara, *Tetrahedron Lett.* p. 2451 (1967).
136. E. J. Smutny, *J. Amer. Chem. Soc.* **89**, 6793 (1967).
137. E. J. Smutny, H. Chung, K. C. Dewhirst, W. Keim, T. M. Shryne, and H. E. Thyret, *Amer. Chem. Soc., Div. Petrol. Chem., Prepr.* **14**, B100 and B112 (1969).
138. S. Takahashi, H. Yamazaki, and N. Hagihara, *Mem. Inst. Sci. Ind. Res., Osako Univ.* **25**, 125 (1968).
139. S. Takahashi, H. Yamazaki, and N. Hagihara, *Bull. Chem. Soc. Jap.* **41**, 254 (1968).
140. S. Takahashi, T. Shibano, and N. Hagihara, *Chem. Commun.* p. 161 (1969).
141. J. F. Kohnle, L. H. Slaugh, and K. L. Nakamaye, *J. Amer. Chem. Soc.* **91**, 5905 (1969).

142. W. E. Walker, R. M. Manyik, K. E. Atkins, and M. L. Farmer, *Tetrahedron Lett.* p. 3817 (1970).
143. G. Wilke, B. Bogdanovic, P. Hardt, P. Heimbach, W. Keim, M. Kröner, W. Oberkirch, K. Tanaka, G. Steinrücke, D. Walter, and H. Zimmerman, *Angew. Chem., Int. Ed. Engl.* **5**, 151 (1966).
144. B. Bogdanovic, P. Heimbach, M. Kröner, G. Wilke, E. G. Hoffmann, and J. Brandt, *Justus Liebigs Ann. Chem.* **727**, 143 (1969); W. Brenner, P. Heimbach, H. Hey, E. W. Muller, and G. Wilke, *ibid.* p. 161.
145. P. Heimbach, P. W. Jolly, and G. Wilke, *Advan. Organometal. Chem.* **8**, 29 (1970).
146. R. Traunmüller, O. E. Polansky, P. Heimbach, and G. Wilke, *Chem. Phys. Lett.* **3**, 300 (1969).
147. R. M. Manyik, W. E. Walker, K. E. Atkins, and E. S. Hammack, *Tetrahedron Lett.* p. 3813 (1970).
148. P. Heimbach, *Angew. Chem., Int. Ed. Engl.* **7**, 882 (1968).
149. N. Yamazaki and S. Murai, *Chem. Commun.* p. 147 (1968).
150. A. J. Canale, W. A. Hewett, T. M. Shryne, and E. A. Youngman, *Chem. Ind. (London)* p. 1054 (1962).
151. A. J. Canale and W. A. Hewett, *J. Polym. Sci. Part B*, **2**, 1041 (1964).
152. G. Allegra, F. Lo Giudice, G. Natta, U. Giannini, G. Fagherazzi, and P. Pino, *Chem. Commun.* p. 1263 (1967).
153. L. S. Meriwether, E. C. Colthup, G. W. Kennerly, and R. N. Reusch, *J. Org. Chem.* **26**, 5155 (1961); L. S. Meriwether, E. C. Colthup, and G. W. Kennerly, *ibid.* p. 5163; E. C. Colthup and L. S. Meriwether, *ibid.* p. 5169; L. S. Meriwether, M. F. Leto, E. C. Colthup, and G. W. Kennerly, *ibid.* **27**, 3930 (1962).
154. V. O. Reikhsfel'd and K. L. Makovetskii, *Russ. Chem. Rev.* **35**, 510 (1966).
155. W. Hübel, *in* "Organic Syntheses via Metal Carbonyls" (I. Wender and P. Pino, eds.), p. 273. Wiley (Interscience), New York, 1968; C. Hoogzand and W. Hübel, *ibid.* p. 343.
156. F. L. Bowden and A. B. P. Lever, *Organometal. Chem. Rev.* **3**, 227 (1968).
157. F. C. Phillips, *Amer. Chem. J.* **16**, 255 (1894).
158. H. Erdmann and O. Makowka, *Chem. Ber.* **37**, 2694 (1904).
159. M. Ziegler and W. Buchholz, *Z. Anal. Chem.* **210**, 344 (1965).
160. M. Ziegler and O. Glemser, *Z. Anal. Chem.* **153**, 247 (1956).
161. O. Makowka, *Chem. Ber.* **41**, 824 (1908).
162. O. N. Temkin, S. M. Brailovskii, R. M. Flid, M. P. Struchkova, V. B. Belyanin, and M. G. Zaitseva, *Kinet. Catal. (USSR)* **5**, 167 (1964).
163. S. M. Brailovskii, O. L. Kaliya, O. N. Temkin, and R. M. Flid, *Kinet. Katal.* **9**, 177 (1968).
164. Y. Odaira, M. Hara, and S. Tsutsumi, *Technol. Rep. Osaka Univ.* **16**, 325 (1965).
165. P. M. Maitlis, J. Graham, and J. Bloodworth, unpublished results (1966).
166. A. Oshima, *Kogyo Kagaku Zasshi* **70**, 1818 (1967).
167. A. V. Babaeva and T. I. Beresneva, *Russ. J. Inorg. Chem.* **11**, 1048 (1966).
168. A. V. Babaeva and T. I. Beresneva, *Russ. J. Inorg. Chem.* **11**, 1434 (1966).
169. A. V. Babaeva and T. I. Beresneva, *Russ. J. Inorg. Chem.* **12**, 1330 (1967).
170. Yu. Ya. Kharitonov, T. I. Beresneva, G. Ya. Mazo, and A. V. Babaeva, *Russ. J. Inorg. Chem.* **12**, 1373 (1967).
171. A. V. Babaeva, T. I. Beresneva, and Yu. Ya. Kharitonov, *Dokl. Akad. Nauk SSSR (English Transl.)* **175**, 647 (1967).
172. Yu. Ya. Kharitonov, A. V. Babaeva, T. I. Beresneva, and G. Ya. Mazo, *Zh. Neorg. Khim.* **13**, 796 (1968).
173. Yu. Ya. Kharitonov, T. I. Beresneva, G. Ya. Mazo, and A. V. Babaeva, *Zh. Neorg. Khim.* **13**, 2184 (1968).

174. L. Malatesta, G. Santarella, L. M. Vallarino, and F. Zingales, *Angew. Chem.* **72**, 34 (1960).
175. A. T. Blomquist and P. M. Maitlis, *J. Amer. Chem. Soc.* **84**, 2329 (1962).
176. P. M. Maitlis and M. L. Games, *Can. J. Chem.* **42**, 183 (1964).
177. P. M. Maitlis, D. Pollock, M. L. Games, and W. J. Pryde, *Can. J. Chem.* **43**, 470 (1965).
178. L. F. Dahl and W. E. Oberhansli, *Inorg. Chem.* **4**, 629 (1965).
179. R. Hüttel and H. J. Neugebauer, *Tetrahedron Lett.* p. 3541 (1964).
180. P. M. Maitlis and D. Pollock, *J. Organometal. Chem.* **26**, 407 (1971).
181. E. Müller, K. Munk, P. Ziemek, and M. Sauerbier, *Justus Liebigs Ann. Chem.* **713**, 40 (1968).
182. T. Hosokawa and I. Moritani, *Tetrahedron Lett.* p. 3021 (1969).
183. M. Avram, I. G. Dinulescu, G. D. Mateescu, E. Avram, and C. D. Nenitzescu, *Rev. Roum. Chim.* **14**, 1181 (1969).
184. F. Zingales, *Ann. Chim. (Rome)* **52**, 1174 (1962).
185. H. Dietl and P. M. Maitlis, *Chem. Commun.* p. 481 (1968).
186. H. Reinheimer, H. Dietl, J. Moffat, D. Wolff, and P. M. Maitlis, *J. Amer. Chem. Soc* **90**, 5321 (1968).
187. H. Dietl, H. Reinheimer, J. Moffat, and P. M. Maitlis, *J. Amer. Chem. Soc.* **92**, 2276 (1970).
188. H. Reinheimer, J. Moffat, and P. M. Maitlis, *J. Amer. Chem. Soc.* **92**, 2285 (1970).
189. G. M. Whitesides and W. J. Ehmann, *J. Amer. Chem. Soc.* **91**, 3800 (1969).
190. M. Avram, E. Avram, G. D. Mateescu, I. G. Dinulescu, F. Chiraleu, and C. D. Nenitzescu, *Chem. Ber.* **102**, 3996 (1969).
191. K. L. Kaiser and P. M. Maitlis, *Chem. Commun.* p. 942 (1971).
192. M. Avram, I. G. Dinulescu, G. D. Mateescu, and C. D. Nenitzescu, *Rev. Roum. Chim.* **14**, 1191 (1969).
193. P. Chini, F. Canziani, and A. Quarta, *Abstr., Inorg. Chim. Acta Symp., 1968* p. D8 (1968).
194. G. Wittig and P. Fritze, *Justus Liebigs Ann. Chem.* **712**, 79 (1968).
195. D. Bryce-Smith, *Chem. Ind. (London)*, p. 239 (1964).
196. K. Moseley and P. M. Maitlis, unpublished results (1970).
197. Y. Takahashi, T. Ito, S. Sakai, and Y. Ishii, *Chem. Commun.* p. 1065 (1970).
198. S. Sakai, Y. Kawashima, Y. Takahashi, and Y. Ishii, *Chem. Commun.* p. 1073 (1967).
199. Y. Kawashima, S. Sakai, Y. Takahashi, and Y. Ishii, *Kogyo Kagaku Zasshi* **72**, 1715 (1969).
200. K. E. Atkins, W. E. Walker, and R. M. Manyik, *Tetrahedron Lett.* p. 3821 (1970).
201. G. Hata, K. Takahashi, and A. Miyake, *Chem. Commun.* p. 1392 (1970).
202. T. Kajimoto, H. Takahashi, and J. Tsuji, *J. Organometal. Chem.* **23**, 275 (1970); Y. Yamamoto and H. Yamazaki, *Bull. Chem. Soc. Jap.* **43**, 2653 (1970).
203. R. F. Heck, *J. Amer. Chem. Soc.* **90**, 5546 (1968).
204. P. M. Maitlis and F. G. A. Stone, *Chem. Ind. (London)* p. 1865 (1962).
205. A. Yamamoto and S. Ikeda, *J. Amer. Chem. Soc.* **89**, 5989 (1967).
206. M. D. Rausch and F. E. Tibbetts, *J. Organometal. Chem.* **21**, 487 (1970).
207. G. Wittig and G. Klar, *Justus Liebigs Ann. Chem.* **704**, 91 (1967).
208. A. Cairncross and W. A. Sheppard, *J. Amer. Chem. Soc.* **90**, 2186 (1968).
209. E. J. Corey and G. H. Posner, *J. Amer. Chem. Soc.* **90**, 5615 (1968).
210. G. M. Whitesides, C. P. Casey, J. S. Filippo, and E. J. Panek, *Trans. N.Y. Acad. Sci.* [2] **29**, 572 (1967).
211. M. Nilsson and O. Wennerström, *Tetrahedron Lett.* p. 3307 (1968).
212. A. Cairncross and W. A. Sheppard, *Abstr. 4th Organometal. Meet., 1969* p. E1 (1969); *J. Amer. Chem. Soc.* **93**, 247 (1971).

213. R. Hüttel and M. Bechter, *Angew. Chem.* **71**, 456 (1959).
214. R. Hüttel, J. Kratzer, and M. Bechter, *Chem. Ber.* **94**, 766 (1961).
215. H. C. Volger, *Amer. Chem. Soc., Div. Petrol. Chem., Prepr.* **14**, B72 (1969); *Rec. Trav. Chim. Pays-Bas* **87**, 225 (1969).
216. H. C. Volger, *Rec. Trav. Chim. Pays-Bas* **86**, 677 (1967).
217. Shell Internationale Research Maatschappij N.V., Neth. Pat., 6,409,545 (1966); *Chem. Abstr.* **65**, 3756e (1966).
218. C. F. Kohll and R. van Helden, *Rec. Trav. Chim. Pays-Bas* **86**, 193 (1967).
219. R. van Helden and G. Verberg, *Rec. Trav. Chim. Pays-Bas* **84**, 1263 (1965).
220. A. F. Ellis (Gulf Research and Development Co.), U.S. Pat., 3,294,484 (1966); *Chem. Abstr.* **66**, 104799w (1967).
221. M. O. Unger and R. A. Fouty, *J. Org. Chem.* **34**, 18 (1969).
222. D. R. Bryant, J. E. McKeon, and B. C. Ream, *Tetrahedron Lett.* p. 3371 (1968).
223. J. M. Davidson and C. Triggs, *Chem. Ind. (London)* p. 1361 (1967).
224. J. M. Davidson and C. Triggs, *J. Chem. Soc., A* p. 1324 (1968).
225. R. G. Brown, J. M. Davidson, and C. Triggs, *Amer. Chem. Soc., Div. Petrol. Chem. Prepr.* **14**, B23 (1969).
226. A. C. Cope and R. W. Siekman, *J. Amer. Chem. Soc.* **87**, 3272 (1965).
227. H. C. Brown and C. W. McGarry, *J. Amer. Chem. Soc.* **77**, 2300, 2306 (1955).
228. T. Sakakibara, S. Nishimura, Y. Odaira, and T. Yoshida, *Tetrahedron Lett.* p. 1019 (1969).
229. T. Sakakibara, T. Teramoto, and Y. Odaira, *Chem. Commun.* p. 1563 (1970).
230. S. Nishimura, Y. Sakakibara, and Y. Odaira, *Chem. Commun.* p. 313 (1969).
231. T. Sakakibara, S. Nishimura, K. Kimura, I. Minato, and Y. Odaira, *J. Org. Chem.* **35**, 3884 (1970).
232. I. Moritani and Y. Fujiwara, *Tetrahedron Lett.* p. 1119 (1967).
233. Y. Fujiwara, I. Moritani, M. Matsuda, and S. Teranishi, *Tetrahedron Lett.* p. 633 (1968).
234. Y. Fujiwara, I. Moritani, M. Matsuda, and S. Teranishi, *Tetrahedron Lett.* p. 3863 (1968).
235. Y. Fujiwara, I. Moritani, and M. Matsuda, *Tetrahedron* **24**, 4819 (1968).
236. I. Moritani, Y. Fujiwara, and S. Teranishi, *Amer. Chem. Soc., Div. Petrol. Chem. Prepr.* **14**, B172 (1969).
237. Y. Fujiwara, I. Moritani, S. Danno, R. Asano, and S. Teranishi, *J. Amer. Chem. Soc.* **91**, 7166 (1969).
238. S. Danno, I. Moritani, and Y. Fujiwara, *Tetrahedron* **25**, 4809 (1969).
239. Y. Fujiwara, I. Moritani, R. Asano, H. Tanaka, and S. Teranishi, *Tetrahedron* **25**, 4815 (1969).
240. S. Danno, I. Moritani, and Y. Fujiwara, *Tetrahedron* **25**, 4819 (1969).
241. Y. Fujiwara, I. Moritani, K. Ikegami, R. Tanaka, and S. Teranishi, *Bull. Chem. Soc. Jap.* **43**, 863 (1970).
242. S. Danno, I. Moritani, and Y. Fujiwara, *J. Chem. Soc. (B)* p. 196 (1971).
243. R. Breslow, "Organic Reaction Mechanisms," p. 155. Benjamin, New York, 1965.
244. G. Olah, S. Kuhn, and S. Flood, *J. Amer. Chem. Soc.* **83**, 4571 (1961); **84**, 3687 (1962); C. H. Ritchie and H. Win, *J. Org. Chem.* **29**, 3093 (1964).
245. R. Long and B. A. Marples (ICI Ltd.), Brit. Pat., 987,516 (1965); *Chem. Abstr.* **63**, 1743a (1965).
246. N. Calderon, E. A. Ofstead, J. P. Ward, W. A. Judy, and K. W. Scott, *J. Amer. Chem. Soc.* **90**, 4133 (1968).
247. R. K. Armstrong, *J. Org. Chem.* **31**, 618 (1966).

248. A. D. Walsh, *Trans. Faraday Soc.* **45**, 179 (1949).
249. R. Hüttel and H. Schmid, *Chem. Ber.* **101**, 252 (1968).
250. B. F. G. Johnson, J. Lewis, and D. A. White, *J. Chem. Soc.* (*A*) 1738 (1970).
251. J. Smidt and R. Sieber, *Angew. Chem.* **71**, 626 (1959).
252. J. Tsuji, K. Ohno, and T. Kajimoto, *Tetrahedron Lett.* p. 4565 (1965).
253. K. Ohno and J. Tsuji, *J. Amer. Chem. Soc.* **90**, 99 (1968).
254. N. E. Hoffman and T. Puthenpurackal, *J. Org. Chem.* **30**, 420 (1965).
255. J. W. Wilt and V. P. Abegg, *J. Org. Chem.* **33**, 923 (1968).
256. J. C. Trebellas, J. R. Olechowski, and H. B. Jonassen, *J. Organometal. Chem.* **6**, 412 (1966).
257. H. Frye and D. Chinn, *Inorg. Nucl. Chem. Lett.* **5**, 613 (1969).
258. E. Vedejs, *J. Amer. Chem. Soc.* **90**, 4751 (1968).
259. H. Dietl and P. M. Maitlis, *Chem. Commun.* p. 759 (1967).
260. L. Cassar, P. E. Eaton, and J. Halpern, *J. Amer. Chem. Soc.* **92**, 6366 (1970).
261. H. Hogeveen and H. C. Volger, *J. Amer. Chem. Soc.* **89**, 2486 (1967); *Chem. Commun.* p. 1133 (1967); *Rec. Trav. Chim. Pays-Bas* **86**, 830 (1967).
262. F. D. Mango and J. Schachtschneider, *J. Amer. Chem. Soc.* **89**, 2484 (1967); **93**, 1123 (1971); F. D. Mango, *Tetrahedron Lett.* p. 4813 (1969).
263. W. Merk and R. Pettit, *J. Amer. Chem. Soc.* **89**, 4788 (1967); R. Pettit, H. Sugahara, J. Wristers, and W. Merk, *Discuss. Faraday Soc.* **47**, 71 (1969).
264. T. J. Katz and S. Cerefice, *Tetrahedron Lett.* p. 2509 (1969); *J. Amer. Chem. Soc.* **91**; 2405 and 6519 (1969); L. Cassar, P. E. Eaton, and J. Halpern, *ibid.* **92**, 3515 (1970). L. Cassar and J. Halpern, *Chem. Commun.* p. 1082 (1970).
265. K. L. Kaiser, R. F. Childs, and P. M. Maitlis, *J. Amer. Chem. Soc.* **93**, 1270 (1971).
266. R. C. Fahey, *Top. Stereochem.* **3**, 237 (1969).
267. J. Chatt, L. M. Vallarino, and L. M. Venanzi, *J. Chem. Soc., London* p. 3413 (1957).
268. J. Chatt, L. M. Vallarino, and L. M. Venanzi, *J. Chem. Soc., London* p. 2496 (1957).
269. A. Panunzi, A. De Renzi, and G. Paiaro, *J. Amer. Chem. Soc.* **92**, 3488 (1970).
270. C. W. Capp, G. W. Godin, R. F. Neale, J. B. Williamson, and B. W. Harris (Distillers Co. Ltd.), Brit. Pat., 918,062 (1963); *Chem. Abstr.* **59**, 5021g (1963).
271. G. Szonyi, *Advan. Chem. Ser.* **70**, 53 (1968).
272. H. Krekeler and H. Schmitz, *Chem.-Ing.-Tech.* **40**, 785 (1968).
273. S. Nakamura and T. Yasui, *J. Catal.* **17**, 366 (1970).
274. M. Tanaka and T. Yasui (Kurashiki Rayon Co.), Jap. Pat., 12127/69; *Chem. Abstr.* **71**, 80671y (1969).
275. M. Tamura, personal communication (1970).
276. British Celanese Ltd., Fr. Pat., 1,423,314 (1966); *Chem. Abstr.* **65**, 13546b (1966).
277. I. I. Moiseev, A. A. Grigor'ev and M. Ya Klimenko, U.S.S.R. Pat., 165,436 (1964); *Chem. Abstr.* **62**, 6396a (1965).
278. Japan Synthetic Chem. Ind. Co. Ltd., Belg. Pat., 628,848 (1963); *Chem. Abstr.* **60**, 7917g (1964).
279. Farbwerke Hoechst, Belg. Pat., 626,669 (1963); *Chem. Abstr.* **60**, 9149c (1964).
280. Shell Internationale Research Maatschappij N.V., Neth. Pat., 6,406,180 (1965); *Chem. Abstr.* **64**, 11091d (1966).
281. Distillers Co. Ltd., Neth. Pat., 6,511,272 (1966); *Chem. Abstr.* **65**, 2130a (1966).
282. ICI Ltd., Neth. Pat., 6,510,921 (1966); *Chem. Abstr.* **65**, 3753g (1966).
283. H. R. Gerberich and W. K. Hall, *Nature* (*London*) **213**, 1120 (1967).
284. ICI Ltd., Brit. Pat., 1,130,760 (1968); *Platinum Metals Rev.* **13**, 43 (1969).
285. R. van Helden, C. F. Kohll, D. Medema, G. Verberg, and T. Jonkhoff, *Rec. Trav. Chim. Pays-Bas* **87**, 961 (1968).

286. W. H. Clement and C. M. Selwitz, *Tetrahedron Lett.* p. 1081 (1962).
287. R. G. Schultz and P. R. Rony, *Amer. Chem. Soc., Div. Petrol. Chem., Prepr.* **12**, 139 (1967).
288. R. G. Schultz and P. R. Rony, *J. Catal.* **16**, 133 (1970).
289. Kurashiki Rayon Co. Ltd., Brit. Pat., 1,142,250 (1969); *Chem. Abstr.* **70**, 96171a (1969).
290. D. Clark, P. Hayden, W. D. Walsh, and W. E. Jones (ICI Ltd.), Brit. Pat., 964,001 (1964); *Chem. Abstr.* **61**, 13199h (1964).
291. P. Hayden (ICI Ltd.), Brit. Pat., 969,162 (1964); *Chem. Abstr.* **62**, 9018e (1965).
292. ICI Ltd., Belg. Pat., 634,595 (1964); *Chem. Abstr.* **60**, 14394c (1964).
293. M. Tamura and T. Yasui, *Kogyo Kagaku Zasshi* **72**, 561 (1969).
294. I. I. Moiseev and M. N. Vargaftik, *Izv. Akad. Nauk SSSR* (*Engl. Transl.*) p. 744 (1965).
295. M. Tamura and T. Yasui, *Chem. Commun.* p. 1209 (1968); Brit. Pat., 1,124,862 (1968); *Chem. Abstr.* **69**, 86391n (1968).
296. M. Tamura and T. Yasui, *Kogyo Kagaku Zasshi* **72**, 575 (1969).
297. M. Tamura and T. Yasui, *Kogyo Kagaku Zasshi* **72**, 578 (1969).
298. M. Tamura and T. Yasui, *Kogyo Kagaku Zasshi* **72**, 581 (1969).
299. M. Tamura, M. Tsutsumi, and T. Yasui, *Kogyo Kagaku Zasshi* **72**, 585 (1969).
300. P. M. Henry, *J. Org. Chem.* **32**, 2575 (1967).
301. E. I. du Pont de Nemours and Co., Brit. Pat., 1,058,995 (1967); *Chem. Abstr.* **66**, 94702t (1967).
302. M. Tamura and T. Yasui, *Kogyo Kagaku Zasshi* **72**, 568 (1969).
303. I. I. Moiseev, M. N. Vargaftik, and Y. K. Syrkin, *Dokl. Akad. Nauk SSSR* (*Engl. Transl.*) **133**, 801 (1960).
304. E. W. Stern and M. L. Spector, *Proc. Chem. Soc., London* p. 370 (1961).
305. A. D. Ketley and J. A. Braatz, *Chem. Commun.* p. 169 (1968).
306. A. D. Ketley and L. P. Fisher, *J. Organometal. Chem.* **13**, 243 (1968).
307. Rhone-Poulenc SA, Fr. Pat., 1,339,614 (1963); *Chem. Abstr.* **60**, 4012g (1964).
308. Shell Internationale Research Maatschappij N.V., Neth. Pat., 301,519 (1965); *Chem. Abstr.* **64**, 6502e (1966).
309. J. S. Anderson, *J. Chem. Soc., London* p. 971 (1934); p. 1042 (1936).
310. A. Aguilo, *Advan. Organometal. Chem.* **5**, 321 (1967).
311. P. M. Henry, *Advan. Chem. Ser.* **70**, 126 (1968).
312. E. W. Stern, *Catal. Rev.* **1**, 73 (1967).
313. I. I. Moiseev, *Amer. Chem. Soc., Div. Petrol. Chem., Prepr.* **14**, B49 (1969); see also *Kinet. Katal.* **11**, 342 (1970).
314. K. Teramoto, T. Oga, S. Kikuchi, and M. Ito, *Yuki Gosei Kagaku Kyokai Shi* **21**, 298 (1963); *Chem. Abstr.* **59**, 7339g (1963).
315. I. I. Moiseev, M. N. Vargaftik, and Ya. K. Syrkin, *Dokl. Akad. Nauk SSSR* **152**, 147 (1963).
316. S. V. Pestrikov, I. I. Moiseev, and L. M. Sverzh, *Russ. J. Inorg. Chem.* **11**, 1173 (1966).
317. P. M. Henry, *J. Amer. Chem. Soc.* **86**, 3246 (1964).
318. P. M. Henry, *J. Amer. Chem. Soc.* **88**, 1595 (1966).
319. W. Hafner, R. Jira, J. Sedlmeier, and J. Smidt, *Chem. Ber.* **95**, 1575 (1962).
320. A. V. Nikiforova, I. I. Moiseev, and Ya. K. Syrkin, *Zh. Obshch. Khim.* **33**, 3239 (1963).
321. R. Jira, J. Sedlmeier, and J. Smidt, *Justus Liebigs Ann. Chem.* **693**, 99 (1966).
322. M. N. Vargaftik, I. I. Moiseev, and Ya. K. Syrkin, *Dokl. Akad. Nauk SSSR* (*Engl. Transl.*) **147**, 804 (1962).
323. I. I. Moiseev, M. N. Vargaftik, and Ya. K. Syrkin, *Izv. Akad Nauk SSSR* (*Engl. Transl.*) p. 1050 (1963).
324. I. I. Moiseev and M. N. Vargaftik, *Dokl. Akad. Nauk SSSR* (*Engl. Transl.*) **166**, 81 (1966).

325. P. M. Henry, personal communication (1970).
326. B. L. Shaw, *Chem. Commun.* p. 464 (1968).
327. I. I. Moiseev, M. N. Vargaftik, S. V. Pestrikov, O. G. Levanda, T. N. Romanova, and Ya. K. Syrkin, *Dokl. Akad. Nauk SSSR* (*Engl. Transl.*) **171**, 813 (1966).
327a. P. M. Henry and O. W. Marks, *Inorg. Chem.* **10**, 373 (1971).
328. W. Kitching, *Organometal. Chem. Rev.* **3**, 35 and 61 (1968).
329. P. M. Henry, *J. Amer. Chem. Soc.* **87**, 990 and 4423 (1965); **88**, 1597 (1966).
330. K. I. Matveev, A. M. Osipov, V. F. Odyakov, Yu. U. Suzdal'nitskaya, I. F. Bukhtoyarov, and O. A. Emel'yanova, *Kinet. Katal.* **3**, 661 (1962).
331. K. I. Matveev, I. F. Bukhtoyarov, N. N. Shul'ts, and O. A. Emel'yanova, *Kinet. Catal.* (*USSR*) **5**, 572 (1964).
332. P. François, *Ann. Chim.* (*Paris*) [14] **4**, 371 (1969).
333. R. Hüttel, H. Dietl, and H. Christ, *Chem. Ber.* **97**, 2037 (1964).
334. J. Smidt, R. Sieber, W. Hafner, and R. Jira (Consortium für Elektrochemische Industrie G.m.b.H), Ger. Pat., 1,176,141 (1964); *Chem. Abstr.* **62**, 447e (1965).
335. H. Okada and H. Hashimoto, *Kogyo Kagaku Zasshi* **69**, 2137 (1966).
336. H. Okada, T. Noma, Y. Katsuyama, and H. Hashimoto, *Bull. Chem. Soc. Jap.* **41**, 1395 (1968).
337. R. Hüttel and H. Christ, *Chem. Ber.* **97**, 1439 (1964).
338. M. N. Vargaftik, I. I. Moiseev, and Ya. K. Syrkin, *Dokl. Akad. Nauk SSSR* **139**, 1396 (1961).
339. I. I. Moiseev, M. N. Vargaftik, and Ya. K. Syrkin, *Dokl. Akad. Nauk SSSR* (*Engl. Transl.*) **130**, 115 (1960).
340. T. Dozono and T. Shiba, *Bull. Jap. Petrol. Inst.* **5**, 8 (1963); *Chem. Abstr.* **59**, 5829b (1963).
341. C. B. Cotterill and F. Dean (ICI Ltd.), Brit. Pat., 941,951 (1963); *Chem. Abstr.* **60**, 4011g (1964).
342. Japan Oil Co., Brit. Pat., 960,195 (1964); *Chem. Abstr.* **61**, 6922f (1964).
343. D. R. Bryant, J. E. McKeon, and P. S. Starcher (Union Carbide Corp.), Fr. Pat., 1,395,129 (1965); *Chem. Abstr.* **63**, 9816g (1965).
344. D. R. Bryant, J. E. McKeon, and P. S. Starcher (Union Carbide Corp.), Belg. Pat., 646,482 (1964); *Chem. Abstr.* **63**, 17906b (1965).
345. Consortium für elektrochemische Industrie GmbH., Brit. Pat., 884,962 (1961); *Chem. Abstr.* **59**, 5024h (1963).
346. C. W. Hargis and H. S. Young (Eastman Kodak Co.), Fr. Pat., 1,359,141 (1964); *Chem. Abstr.* **61**, 13196f (1964).
347. Farbwerke Hoechst AG, Belg. Pat., 648,304 (1964); *Chem. Abstr.* **63**, 13082h (1965).
348. H. J. Hagemeyer, F. C. Canter, and H. F. Goss (Eastman Kodak Co.), Fr. Pat., 1,421,181 (1965); *Chem. Abstr.* **65**, 13549b (1966).
349. J. Smidt, W. Hafner, and R. Jira (Consortium für Elektrochemische Industrie GmbH.), U.S. Pat., 3,080,425 (1963); *Chem. Abstr.* **59**, 7375g (1963).
350. B. W. Harris (Distillers Co. Ltd.), Fr. Pat., 1,397,054 (1965); *Chem. Abstr.* **63**, 4164b (1965).
351. Distillers Co. Ltd., Neth. Pat., 6,409,616 (1965); *Chem. Abstr.* **63**, 8203f (1965).
352. Badische Aniline u. Soda Fabrik AG., Belg. Pat., 658,801 (1965); *Chem. Abstr.* **64**, 6499b (1966).
353. Celanese Corp. of America, Brit. Pat., 1,047,172 (1966); *Chem. Abstr.* **66**, 18513s (1967).
354. J. Smidt, *Chem. Eng. News* July 8, 1963, p. 50.
355. W. H. Clement and C. M. Selwitz, *J. Org. Chem.* **29**, 241 (1964).

356. R. J. Ouellette and C. Levin, *J. Amer. Chem. Soc.* **90**, 6889 (1968); **93**, 471 (1971).
357. A. P. Belov, I. I. Moiseev, and N. G. Uvarova, *Izv. Akad. Nauk SSSR* (*Engl. Transl.*) p. 2194 (1965).
358. R. Ninomiya, M. Sato, and T. Shiba, *Bull. Jap. Petrol. Inst.* **7**, 31 (1965); *Chem. Abstr.* **63**, 17829f (1965).
359. I. I. Moiseev, A. P. Belov, V. A. Igoshin, and Ya. K. Syrkin, *Dokl. Akad. Nauk SSSR* **173**, 863 (1967).
360. I. I. Moiseev, A. P. Belov, V. A. Igoshin, and Ya. K. Syrkin, *Dokl. Akad. Nauk SSSR* (*Engl. Transl.*) **174**, 256 (1967).
361. D. Clark, P. Hayden, and R. D. Smith, *Discuss. Faraday Soc.* **46**, 98 (1968).
362. D. Clark and P. Hayden, *Amer. Chem. Soc., Div. Petrol. Chem., Prepr.* **11**, D5 (1966).
363. D. Clark, P. Hayden, and R. D. Smith, *Amer. Chem. Soc., Div. Petrol. Chem., Prepr.* **14**, B10 (1969).
364. K. Shimizu and N. Ohta, *Kogyo Kagaku Zasshi* **72**, 1773 (1969).
365. D. R. Bryant and J. E. McKeon, *Amer. Chem. Soc., Div. Petrol. Chem., Prepr.* **14**, B1A (1969).
366. Japan Synthetic Chem. Ind. Co. Ltd., Belg. Pat., 618,071 (1962) *Chem. Abstr.* **59** 12715c (1963).
367. I. I. Moiseev, N. N. Yukhtin, S. S. Bobkov, O. G. Levanda, M. N. Vargaftik, and V. V. Yakshin, U.S.S.R. Pat., 154,537 (1963); *Chem. Abstr.* **60**, 6753d (1964).
368. ICI Ltd., Belg. Pat., 638,268 (1964); *Chem. Abstr.* **62**, 10343b (1965).
369. W. Kroenig and B. Frenz (Farbenfabriken Bayer), Fr. Pat., 1,397,083 (1965); *Chem. Abstr.* **63**, 4210g (1965).
370. R. Jira (Consortium für Elektrochemische Industrie GmbH.), Ger. Pat., 1,190,931 (1965); *Chem. Abstr.* **63**, 6865g (1965).
371. Shell Internationale Research Maatschappij N.V., Neth. Appl., 287,873 (1965); *Chem. Abstr.* **63**, 9820d (1965).
372. Consortium für elektrochemische Industrie GmbH., Ger. Pat., 1,191,362 (1965); *Chem. Abstr.* **63**, 9820h (1965).
373. ICI Ltd., Neth. Pat., 6,412,134 (1965); *Chem. Abstr.* **63**, 13085f (1965).
374. B. W. Harris (Distillers Co. Ltd.), Fr. Pat., 1,412,151 (1965); *Chem. Abstr.* **64**, 3362h (1966).
375. E. I. du Pont de Nemours and Co., Neth. Pat., 6,501,904 (1965); *Chem. Abstr.* **64**, 4948c (1966).
376. B. W. Harris (Distillers Co. Ltd.), Brit. Pat., 1,010,548 (1965); *Chem. Abstr.* **64**, 6502a (1966).
377. D. W. Lum and K. Koch (National Distillers and Chem. Corp.), U.S. Pat., 3,227,747 (1966); *Chem. Abstr.* **64**, 9599d (1966).
378. Celanese Corp. of America, Neth. Pat., 6,510,481 (1966): *Chem. Abstr.* **65**, 619e (1966).
379. E. I. du Pont de Nemours and Co., Neth. Pat., 6,509,320 (1966) *Chem. Abstr.* **65**, 2130b (1966).
380. W. D. Schaeffer (Union Oil Co. of California), U.S. Pat., 3,260,739 (1966); *Chem. Abstr.* **65**, 8767d (1966).
381. W. D. Schaeffer (Union Oil Co. of California), U.S. Pat., 3,277,158 (1966); *Chem. Abstr.* **65**, 20015h (1966).
382. Knapsack AG., Neth. Pat., 6,604,391 (1966); *Chem. Abstr.* **66**, 55053n (1967).
383. ICI Ltd., Neth. Pat., 6,607,927 (1966); *Chem. Abstr.* **66**, 115299j (1967).
384. Knapsack AG., Neth. Pat., 6,608,847 (1966); *Chem. Abstr.* **66**, 115318q (1967).
385. National Distillers and Chemical Corp., Neth. Pat., 6,608,281 (1966); *Chem. Abstr.* **66**, 115321k (1967).

386. A. Aguilo (Celanese Corp. of America), Belg. Pat., 668,284 (1966); *Chem. Abstr.* **65**, 15236h (1966).
387. M. Green, R. N. Haszeldine, and J. Lindley, *J. Organometal. Chem.* **6**, 107 (1966).
388. P. M. Henry, *Abstr. 154th Amer. Chem. Soc. Meet.* p. S78 (1967).
389. C. B. Anderson and S. Winstein, *J. Org. Chem.* **28**, 605 (1963).
390. R. G. Brown and J. M. Davidson, *J. Chem. Soc., A* (1971) (in press).
391. I. I. Moiseev, A. P. Belov, and Ya. K. Syrkin, *Izv. Akad. Nauk SSSR* (*Engl. Transl.*) p. 1395 (1963).
392. A. P. Belov, G. Yu, Pek, and I. I. Moiseev, *Izv. Akad. Nauk SSSR* (*Engl. Transl.*) p. 2170 (1965).
393. E. W. Stern, *Proc. Chem. Soc., London* p. 111 (1963).
394. Consortium für Elektrochemische Industrie GmbH., Fr. Pat., 1,370,867 (1964); *Chem. Abstr.* **62**, 9018d (1965).
395. ICI Ltd., Neth. Pat., 6,501,823 (1965); *Chem. Abstr.* **64**, 4949f (1966).
396. ICI Ltd., Neth. Pat., 6,508,238 (1965); *Chem. Abstr.* **64**, 17430g (1966).
397. W. Kitching, Z. Rappoport, S. Winstein, and W. G. Young, *J. Amer. Chem. Soc.* **88**, 2054 (1966).
398. W. Kroenig and B. Frenz (Farbenfabriken Bayer), Ger. Pat., 1,262,996 (1968); *Chem. Abstr.* **68**, 95342d (1968).
399. H. Arai, S. Kojima, K. Fujimoto, and T. Kunugi, *Kogyo Kagaku Zasshi* **72**, 1767 (1969).
400. ICI Ltd., Belg. Pat., 635,426 (1964); *Chem. Abstr.* **61**, 11896d (1964).
401. A. P. Belov and I. I. Moiseev, *Izv. Akad. Nauk SSSR* (*Engl. Transl.*) p. 114 (1966).
402. R. G. Shultz and D. E. Gross, *Advan. Chem. Ser.* **70**, 97 (1968).
403. O. G. Levanda and I. I. Moiseev, *Zh. Org. Khim.* **4**, 1533 (1968).
404. M. N. Vargaftik, I. I. Moiseev, Ya. K. Syrkin, and V. V. Yakshin, *Izv. Akad. Nauk SSSR* (*Engl. Transl.*) p. 868 (1962).
405. T. Matsuda, T. Mitsuyasu, and Y. Nakamura, *Rogyo Kagaku Zasshi* **72**, 1751 (1969).
406. ICI Ltd., Fr. Pat., 1,374,736 (1964); *Chem. Abstr.* **62**, 3943e (1965).
407. T. Matsuda and Y. Nakamura, *Kogyo Kagaku Zasshi* **72**, 1756 (1969).
408. W. C. Baird, *J. Org. Chem.* **31**, 2411 (1966).
409. M. A. Battiste and J. W. Nebzydoski, *J. Amer. Chem. Soc.* **91**, 6887 (1969).
410. R. Baker, D. E. Halliday, and T. J. Mason, *Tetrahedron Lett.* p. 591 (1970).
411. S. Uemura and K. Ichikawa, *Nippon Kagaku Zasshi* **88**, 893 (1967); *Chem. Abstr.* **68**, 114224k (1968).
412. S. Uemura and K. Ichikawa, *Bull. Chem. Soc. Jap.* **40**, 1016 (1967).
413. ICI Ltd., Brit. Pat., 1,138,366 (1969); *Platinum Metals Rev.* **13**, 83 (1969).
414. Pullman Inc., Brit. Pat., 1,104,805 (1968); *Chem. Abstr.* **68**, 104562e (1968).
415. W. G. Lloyd and B. J. Luberoff, *J. Org. Chem.* **34**, 3949 (1969).
416. J. Smidt, W. Hafner, R. Jira, R. Sieber, T. Sedlmeier, and A. Sabel, *Angew. Chem., Int. Ed. Engl.* **1**, 80 (1962); J. Smidt, *Chem. Ind.* (*London*) p. 54 (1962).
417. J. Smidt and A. Sabel, Consortium für Elektrochemische Industrie GmbH., Ger. Pat., 1,127,888 (1962); *Chem. Abstr.* **57**, 3634c (1962).
418. Shell Internationale Research Maatschappij N.V., Fr. Pat., 1,385,309 (1965); *Chem. Abstr.* **63**, 8206e (1965).
419. Pullman Inc., Brit. Pat., 1,007,815 (1965); *Chem. Abstr.* **64**, 4949d (1966).
420. E. W. Stern, M. L. Spector, and H. P. Leftin, *J. Catal.* **6**, 152 (1966).
421. C. F. Kohll and R. van Helden, *Rec. Trav. Chim. Pays-Bas* **87**, 481 (1968).
422. H. C. Volger, *Rec. Trav. Chim. Pays-Bas* **87**, 501 (1968).
423. ICI Ltd., Neth. Pat., 6,511,468 (1966); *Chem. Abstr.* **65**, 3750e (1966).

424. M. Tamura and T. Yasui (Kurashiki Rayon Co.), Jap. Pat., 08,241/68; *Chem. Abstr.* **70**, 19548b (1969).
425. P. M. Henry, *Amer. Chem. Soc., Div. Petrol. Chem., Prepr.* **14**, B15 (1969).
426. R. Jira, H. Prigge, A. Sabel, and J. Smidt, *Abstr. 4th Organometal. Meet., 1969* p. L5 (1969).
427. A. Sabel, J. Smidt, R. Jira, and H. Prigge, *Chem. Ber.* **102**, 2939 (1969).
428. J. E. McKeon, P. Fitton, and A. A. Griswold, *Tetrahedron* (1971) in press; J. E. McKeon and P. E. Fitton, *ibid.* (1971) in press.
429. M. Tamura and T. Yasui, *Kogyo Kagaku Zasshi* **72**, 572 (1969).
430. M. Yamaji, Y. Fujiwara, T. Imanaka, and S. Teranishi, *Bull. Chem. Soc. Jap.* **43**, 2659 (1970).
431. D. G. Brady, *Chem. Commun.* p. 434 (1970).
432. A. Misono, Y. Uchida, and K. I. Furuhata, *Bull. Chem. Soc. Jap.* **43**, 1243 (1970).
433. Kurashiki Rayon Co. Ltd., Fr. Pat., 1,442,980 (1966); *Chem. Abstr.* **66**. 104707q (1967).
434. K. Shoda and M. Yano (Kurashiki Rayon Co.), Jap. Pat., 22,847/68; *Chem. Abstr.* **70**, 77349y (1969).
435. J. M. Davidson and C. Triggs, *J. Chem. Soc., A* p. 1331 (1968).
436. T. Tissue and W. J. Downs, *Chem. Commun.* p. 410 (1969).
437. D. R. Bryant, J. E. McKeon, and B. C. Ream, *J. Org. Chem.* **33**, 4123 (1968).
438. D. R. Bryant, J. E. McKeon, and B. C. Ream, *J. Org. Chem.* **34**, 1107 (1969).
439. P. Fitton, J. E. McKeon, and B. C. Ream, *Chem. Commun.* p. 370 (1969).
440. Lummus Co., Brit. Pat., 1,141,238 (1969); *Platinum Metals Rev.* **13**, 83 (1969).
441. J. J. Berzelius, *Justus Liebigs. Ann. Chem.* **13**, 435 (1828).
442. W. G. Lloyd, *J. Org. Chem.* **32**, 2816 (1967).
443. H. B. Charman, *Nature (London)* **212**, 278 (1966); *J. Chem. Soc., B* p. 629 (1967).
444. L. Vaska and J. Di Luzio, *J. Amer. Chem. Soc.* **88**, 4100 (1966).
445. E. S. Mayo (E. I. du Pont de Nemours and Co.), U.S. Pat., 3,245,917 (1966); *Chem. Abstr.* **65**, 1290g (1966).
446. A. B. Fasman, V. A. Golodov, and D. V. Sokol'skii, *Dokl. Akad. Nauk SSSR* **155**, 434 (1964).
447. G. D. Zakumbaeva, N. F. Noskova, E. N. Konaev, and D. V. Sokol'skii, *Dokl. Akad. Nauk SSSR (Engl. Transl.)* **156**, 1386 (1964).
448. A. B. Fasman, G. G. Kutyukov, and D. V. Sokolskii, *Dokl. Akad. Nauk SSSR (Engl. Transl.)* **158**, 958 (1964).
449. A. B. Fasman, V. A. Golodov, and D. V. Sokolskii, *Zh. Fiz. Khim.* **38**, 1545 (1964).
450. V. A. Golodov, G. G. Kutyukov, A. B. Fasman, and D. V. Sokolskii, *Russ. J. Inorg. Chem.* **9**, 1257 (1964).
451. A. B. Fasman, V. D. Markov, and D. V. Sokol'skii, *Zh. Prikl. Khim.* **38**, 791 (1965).
452. A. B. Fasman, G. G. Kutyukov, and D. V. Sokol'skii, *Russ. J. Inorg. Chem.* **10**, 727 (1965).
453. G. G. Kutyukov, A. B. Fasman, A. E. Lyuts, Yu. A. Kushnikov, V. F. Vozdvizhenskii, and V. A. Golodov, *Zh. Fiz. Khim.* **40**, 1468 (1966).
454. V. D. Markov, V. A. Golodov, G. G. Kutyukov, and A. B. Fasman, *Zh. Fiz. Khim.* **40**, 1527 (1966).
455. V. A. Golodov, A. B. Fasman, D. V. Sokol'skii, and V. D. Markov, U.S.S.R. Pat., 163,593 (1964); *Chem. Abstr.* **61**, 13933a (1964).
456. J. V. Kingston and G. R. Scollary, *Chem. Commun.* p. 455 (1969).
457. ICI Ltd., Neth. Pat., 6,606,109 (1966); *Chem. Abstr.* **66**, 75689c (1967).
458. A. Aguilo, *J. Catal.* **13**, 283 (1969).
459. ICI Ltd., Neth. Pat., 6,501,760 (1965); *Chem. Abstr.* **64**, 6500b (1966).

460. E. O. Fischer and H. Werner, Ger. Pat., 1,181,708 (1964); *Chem. Abstr.* **62**, 4947b (1965).
461. S. Takahashi, K. Sonogashira, and N. Hagihara, *Mem. Inst. Sci. Ind. Res., Osako Univ.* **23**, 69 (1966).
462. S. Takahashi, K. Sonogashira, and N. Hagihara, *Nippon Kagaku Zasshi* **87**, 610 (1966).
463. S. Otsuka, A. Nakamura, and Y. Tatsuno, *J. Amer. Chem. Soc.* **91**, 6994 (1969).
464. E. W. Stern, *Chem. Commun.* p. 736 (1970).
465. C. J. Nyman, C. E. Wymore, and G. Wilkinson, *J. Chem. Soc., A* p. 561 (1968); P. J. Hayward, D. M. Blake, G. Wilkinson, and C. J. Nyman, *J. Amer. Chem. Soc.* **92**, 5873 (1970).
466. K. Yamamoto, M. Kumada, I. Nakajima, K. Maeda, and N. Imaki, *J. Organometal. Chem.* **13**, 329 (1968).
467. M. Tada, Y. Kuroda, and T. Sato, *Tetrahedron Lett.* p. 2871 (1969).
468. F. Bonati, G. Minghetti, T. Boschi, and B. Crociani, *J. Organometal. Chem.* **25**, 255 (1970); A. Burke, A. L. Balch, and J. H. Enemark, *J. Amer. Chem. Soc.* **92**, 2555 (1970); B. Crociani, T. Boschi, and U. Belluco, *Inorg. Chem.* **9**, 2021 (1970).
469. I. L. Mador and A. U. Blackham (National Distillers and Chem. Corp.), U.S. Pat., 3,114,762 (1963); *Chem. Abstr.* **60**, 9149a (1964).
470. H. Grassner and F. Stolp (Badische Aniline und Soda-Fabrik), Ger. Pat., 1,179,922 (1964); *Chem. Abstr.* **62**, 1572g (1965).
471. K. Oda, N. Yasuoka, T. Ueki, N. Kasai, M. Kakudo, Y. Tezuka, T. Ogura, and S. Kawaguchi, *Chem. Commun.* p. 989 (1968); K. Oda, N. Yasuoka, T. Ueki, N. Kasai, and M. Kakudo, *Bull. Chem. Soc. Jap.* **43**, 362 (1970).
472. S. Okeya, T. Ogura, and S. Kawaguchi, *Kogyo Kagaku Zasshi* **72**, 1656 (1969).
473. T. M. Shryne, E. J. Smutny, and D. P. Stevenson, U.S. Pat. 3,493,617 (1970).
474. L. Turner (B.P. Co. Ltd.), Brit. Pat., 932,748 (1963); *Chem. Abstr.* **60**, 405g (1964).
475. R. Cramer, *J. Amer. Chem. Soc.* **88**, 2272 (1966).
476. H. A. Tayim and J. C. Bailar, *J. Amer. Chem. Soc.* **89**, 3420 (1967).
477. R. Cramer and R. V. Lindsey, *J. Amer. Chem. Soc.* **88**, 3534 (1966).
478. I. I. Moiseev and S. V. Pestrikov, *Dokl. Akad. Nauk SSSR* **171**, 722 (1966).
479. I. I. Moiseev and S. V. Pestrikov, *Izv. Akad. Nauk SSSR (Engl. Transl.)* p. 1690 (1965).
480. I. I. Moiseev, S. V. Pestrikov, and L. M. Sverzh, *Izv. Akad. Nauk SSSR* p. 1866 (1966).
481. I. I. Moiseev, A. A. Grigor'ev, and S. V. Pestrikov, *Zh. Org. Khim.* **4**, 354 (1968).
482. S. V. Pestrikov, I. I. Moiseev, and L. M. Sverzh, *Kinet. Katal.* **10**, 74 (1969).
483. I. I. Moiseev, M. N. Vargaftik, and Ya. K. Syrkin, *Dokl. Akad. Nauk SSSR (Engl. Transl.)* **153**, 960 (1963).
484. I. I. Moiseev and A. A. Grigor'ev, *Dokl. Akad. Nauk SSSR (Engl. Transl.)* **178**, 132 (1968).
485. M. Donati and F. Conti, *Inorg. Nucl. Chem. Lett.* **2**, 343 (1966).
486. M. Donati and F. Conti, *Tetrahedron Lett.* p. 4953 (1966).
487. D. Morelli, R. Ugo, F. Conti, and M. Donati, *Chem. Commun.* p. 801 (1967).
488. G. Pregaglia, M. Donati, and F. Conti, *Chim. Ind. (Milan)* **49**, 1277 (1867).
489. G. C. Bond and M. Hellier, *J. Catal.* **4**, 1 (1965).
490. M. B. Sparke, L. Turner, and A. J. M. Wenham, *J. Catal.* **4**, 332 (1965).
491. F. Asinger, B. Fell, and P. Krings, *Tetrahedron Lett.* p. 633 (1966).
492. N. R. Davies, *Aust. J. Chem.* **17**, 212 (1964).
493. N. R. Davies, *Nature (London)* **201**, 490 (1964).
494. B. Cruikshank and N. R. Davies, *Aust. J. Chem.* **19**, 815 (1966).
495. N. R. Davies, A. D. Di Michiel, and V. A. Pickles, *Aust. J. Chem.* **21**, 385 (1968)
496. J. F. Harrod and A. J. Chalk, *J. Amer. Chem. Soc.* **86**, 1776 (1964).

497. S. Carrà and V. Ragaini, *J. Catal.* **10**, 230 (1968); V. Ragaini, G. Somerzi, and S. Carrà, *ibid.* **13**, 20 (1969).
498. J. F. Harrod and A. J. Chalk, *J. Amer. Chem. Soc.* **88**, 3491 (1966).
499. J. F. Harrod and A. J. Chalk, *Nature* (*London*) **205**, 280 (1965).
500. N. R. Davies, *Rev. Pure Appl. Chem.* **17**, 83 (1967).
501. H. Alper, P. C. LePort, and S. Wolfe, *J. Amer. Chem. Soc.* **91**, 7554 (1969).
502. H. Boennemann, *Angew. Chem., Int. Ed. Engl.* **9**, 736 (1970).
503. F. G. Cowherd and J. L. von Rosenberg, *J. Amer. Chem. Soc.* **91**, 2158 (1969); W. T. Hendrix, F. G. Cowherd, and J. L. von Rosenberg, *Chem. Commun.* p. 97 (1968).
504. A. Marbach and Y. L. Pascal, *C. R. Acad. Sci., Ser. C* **268**, 1074 (1969).
505. H. A. Tayim and A. Vassilian, *Chem. Commun.* p. 630 (1970).
506. D. Wittenberg and H. Seibt (Badische Aniline u. Soda-Fabrik AG.), Ger. Pat., 1,136,329 (1962); *Chem. Abstr.* **58**, 4442b (1963).
507. E. A. Zuech (Phillips Petroleum Co.), U.S. Pat., 3,387,045 (1968); *Chem. Abstr.* **69**, 58871a (1968).
508. M. Misono, Y. Saito, and Y. Yoneda, *J. Catal.* **10**, 200 (1968).
509. A. D. Ketley, J. A. Braatz, J. Craig, and R. Cole, *Amer. Chem. Soc., Div. Petrol. Chem., Prepr.* **14**, B142 (1969); A. D. Ketley, J. A. Braatz, and J. Craig, *Chem. Commun.* p. 1117 (1970).
510. H. Frye, E. Kuljian, and J. Viebrock, *Inorg. Nucl. Chem. Lett.* **2**, 119 (1966).
511. H. Itatani and J. C. Bailar, *J. Amer. Oil Chem. Soc.* **44**, 147 (1967).
512. J. Lukas, S. Coren, and J. E. Blom, *Chem. Commun.* p. 1303 (1969).
513. F. J. Karol and W. L. Carrick (Union Carbide Corp.), U.S. Pat., 3,287,427 (1966); *Chem. Abstr.* **66**, 115359d (1967).
514. N. D. Zelinsky and G. S. Pavlov, *Chem. Ber.* **66**, 1420 (1933).
515. S. Carrà, P. Beltrame, and V. Ragaini, *J. Catal.* **3**, 353 (1964).
516. S. Carrà and V. Ragaini, *Tetrahedron Lett.* p. 1079 (1967).
517. T. A. Schenach and F. F. Caserio, *J. Organometal. Chem.* **18**, P17 (1969).
518. R. W. Howsam and F. J. McQuillin, *Tetrahedron Lett.* p. 3667 (1968); K. Dunne and F. J. McQuillin, *J. Chem. Soc.* (*C*) p. 2196 and 2200 (1970).
519. I. T. Harrison, E. Kimura, E. Bohme, and J. H. Fried, *Tetrahedron Lett.* p. 1589 (1969).
520. M. Pesez and J. F. Burtin, *Bull. Soc. Chim. Fr.* [5] p. 1996 (1959).
521. P. N. Rylander, N. Himelstein, D. R. Steele, and T. Kreidl, *Engelhard Ind., Tech. Bull.* **3**, 61 (1962).
522. E. B. Maxted and S. M. Ismail, *J. Chem. Soc., London* p. 1750 (1964).
523. M. F. Sloan, A. S. Matlack, and D. S. Breslow, *J. Amer. Chem. Soc.* **85**, 4014 (1963).
524. R. F. Heck (Hercules Powder Co.), U.S. Pat., 3,270,087 (1966); *Chem. Abstr.* **65**, 16857d (1966).
525. ICI Ltd., Neth. Pat., 6,611,373 (1967); *Chem. Abstr.* **67**, 54266t (1967).
526. J. C. Bailar and H. Itatani, *J. Amer. Chem. Soc.* **89**, 1592 (1967).
527. I. Jardine and F. J. McQuillin, *Tetrahedron Lett.* p. 4871 (1966).
528. H. Itatani and J. C. Bailar, *J. Amer. Chem. Soc.* **89**, 1600 (1967).
529. E. N. Frankel, E. A. Emken, H. Itatani, and J. C. Bailar, *J. Org. Chem.* **32**, 1447 (1967).
530. R. W. Adams, G. E. Batley, and J. C. Bailar, *J. Amer. Chem. Soc.* **90**, 6051 (1968).
531. R. W. Adams, G. E. Batley, and J. C. Bailar, *Inorg. Nucl. Chem. Lett.* **4**, 455 (1968).
531a. G. Raper and W. S. McDonald, *Chem. Commun.* p. 655 (1970).
531b. M. Giustiniani, G. Dolcetti, R. Pietropaolo, and U. Belluco, *Inorg. Chem.* **8**, 1048 (1969).
531c. R. L. Golden and E. I. Korchak (Halcon International Inc.), Brit. Pat., 1,122,702 (1968); *Platinum Metals Rev.* **13**, 43 (1969).

532. H. Christ and R. Hüttel, *Angew. Chem.* **75**, 921 (1963).
533. P. Mushak and M. A. Battiste, *J. Organometal. Chem.* **17**, P46 (1969).
534. J. V. Kingston and G. R. Scollary, *Chem. Commun.* p. 362 (1970).
535. B. L. Shaw, *Chem. Ind. (London)* p. 1190 (1962).
536. S. D. Robinson and B. L. Shaw, *J. Chem. Soc., London* p. 4806 (1963).
537. M. Donati and F. Conti, *Tetrahedron Lett.* p. 1219 (1966).
538. A. D. Ketley and J. A. Braatz, *J. Organometal. Chem.* **9**, P5 (1967).
539. R. G. Schultz, *Tetrahedron Lett.* p. 301 (1964); *Tetrahedron* **20**, 2809 (1964).
540. R. G. Schultz (Monsanto Co.), U.S. Pat., 3,369,035 (1968); *Chem. Abstr.* **68**, 95986s (1968).
541. M. S. Lupin and B. L. Shaw, *Tetrahedron Lett.* p. 15 (1964).
542. M. S. Lupin, J. Powell, and B. L. Shaw, *J. Chem. Soc., A* p. 1687 (1966)
543. T. Yukawa and S. Tsutsumi, *Inorg. Chem.* **7**, 1458 (1968).
544. S. M. Brailovskii, O. L. Kaliya, O. N. Temkin, and R. M. Flid, *Kinet. Katal.* **9**, 177 (1968).
545. R. Hüttel and J. Kratzer, *Angew. Chem.* **71**, 456 (1959).
546. L. M. Zaitsev, A. P. Belov, M. N. Vargaftik, and I. I. Moiseev, *Russ. J. Inorg. Chem.* **12**, 203 (1967).
547. H. A. Tayim, *Chem. Ind. (London)* p. 1468 (1970).
548. H. Hachenberg, G. Man, F. Merter, and U. Schwenk (Farbwerke Hoechst AG.), Ger. Pat., 1,202,263 (1965); *Chem. Abstr.* **64**, 3355f (1966).
549. J. Smidt and H. Krekeler, *Erdoel Kohle* **16**, 560 (1963).
550. Farbwerke Hoechst AG., Brit. Pat., 982,643 (1965); *Chem. Abstr.* **63**, 9816e (1965).
551. Farbwerke Hoechst AB., Neth. Pat., 6,504,213 (1965); *Chem. Abstr.* **64**, 8030d (1966).
552. D. R. Fahey, *Chem. Commun.* p. 417 (1970); *J. Organometal. Chem.* **27**, 283 (1971).
553. ICI Ltd., Brit. Pat., 1,159,950 (1969); *Platinum Metals Rev.* **14**, 39 (1970).
554. Shell Internationale Research Maatschappij, Neth. Pat., 6,503,362 (1966); *Chem. Abstr.* **66**, 65634s (1967); additional to Neth. Pat., 6,408,476 (1965); *Chem. Abstr.* **63**, 499c (1965).
555. G. Paiaro, A. De Renzi, and R. Palumbo, *Chem. Commun.* p. 1150 (1967).
556. R. Palumbo, A. De Renzi, A. Panunzi, and G. Paiaro, *J. Amer. Chem. Soc.* **91**, 3874 (1969); A. Panunzi, A. De Renvi, R. Palumbo, and G. Paiaro, *ibid.* p. 3879.
557. R. N. Haszeldine, R. V. Parish and D. W. Robbins, *J. Organometal. Chem.* **23**, C33 (1970).
558. J. E. McKeon and P. S. Starcher (Union Carbide Corp.), U.S. Pat., 3,318,906 (1967); *Chem. Abstr.* **68**, 39466k (1968).
559. H. Hirai, H. Sawai, and S. Makishima, *Bull. Chem. Soc. Jap.* **43**, 1148 (1970); H. Hirai and H. Sawai, *ibid.* p. 2208.
560. ICI Ltd., Brit. Pat., 1,142,991 (1969); *Platinum Metals Rev.* **13**, 123 (1969).
561. W. Beck, M. Bauder, W. P. Fehlhemmer, P. Pöllman, and H. Schachl, *Inorg. Nucl. Chem. Lett.* **4**, 143 (1968).
562. J. Tsuji and N. Iwamoto, *Chem. Commun.* p. 828 (1966).
563. E. W. Stern and M. L. Spector, *J. Org. Chem.* **31**, 596 (1966).
564. P. Haynes, L. H. Slaugh, and J. F. Kohnle, *Tetrahedron Lett.* p. 365 (1970).
565. ICI Ltd., Neth. Pat., 6,603,612 (1966); *Chem. Abstr.* **66**, 28511d (1967).
566. B. Braithwaite and D. Wright, *Chem. Commun.* p. 1329 (1969).
567. S. O'Brien, *J. Chem. Soc., A* p. 9 (1970).
568. L. H. Sommer and J. E. Lyons, *J. Amer. Chem. Soc.* **91**, 7061 (1969).
569. G. C. Bond, "Catalysis by Metals." Academic Press, New York, 1962.

570. J. W. Colton and J. E. Jewett, Jr. (Scientific Design Co. Inc.), Belg. Pat., 623,399 (1963); *Chem. Abstr.* **60**, 9154a (1964).
571. A. N. Naglieri (Scientific Design Co. Inc.), Belg. Pat., 623,229 (1963); *Chem. Abstr.* **60**, 9153h (1964).
572. A. Mitsutani, K. Tanaka, and M. Yano, *Kogyo Kagaku Zasshi* **68**, 1219 (1965).
573. Farbwerke Hoechst AG., Belg. Pat., 662,624 (1965); *Chem. Abstr.* **65**, 3752b (1966).
574. F. Wattimena (Shell Oil Co.), U.S. Pat., 3,293,291; (1966); *Chem. Abstr.* **66**, 65100q (1967).
575. W. Schwerdtel, *Chem. Ind. (London)* p. 1559 (1968).
576. C. Kemball and W. R. Patterson, *Proc. Roy. Soc., Ser. A* **270**, 219 (1962); W. R. Patterson and C. Kemball, *J. Catal.* **2**, 465 (1963).
577. A. R. Blake, J. G. Sunderland, and A. T. Kuhn, *J. Chem. Soc., A* p. 3015 (1969).
578. H. R. Gerberich, N. W. Cant, and W. K. Hall, *J. Catal.* **16**, 204 (1970); N. W. Cant and W. K. Hall, *ibid.* p. 220.
579. Knapsack, A. G., Brit. Pat., 1,107,495 (1968); *Chem. Abstr.* **68**, 114086s (1968).
580. L. Crombie, P. A. Jenkins, D. A. Mitchard, and J. C. Williams, *Tetrahedron Lett.* p. 4297 (1967).
581. J. R. Lacher, A. Kianpour, F. Oetting, and J. D. Park, *Trans. Faraday Soc.* **52**, 1500 (1956).
582. R. L. Burwell, *Accounts Chem. Res.* **2**, 289 (1969); *Chem. Eng. News* Aug. 22 (1966), p. 56.
583. J. J. Rooney, *Chem. Brit.* p. 242 (1966).
584. H. A. Quinn, M. A. McKervey, W. R. Jackson, and J. J. Rooney, *J. Amer. Chem. Soc.* **92**, 2922 (1970).
585. J. S. Matthews, D. C. Ketter, and R. F. Hall, *J. Org. Chem.* **35**, 1694 (1970).
586. Farbwerke Hoechst AG., Neth. Pat., 6,507,073 (1965); *Chem. Abstr.* **64**, 17430f (1966).
587. S. Beesley and V. F. G. Cooke (British Celanese), Brit. Pat., 1,120,469 (1968); *Chem. Abstr.* **69**, 51633r (1968).
588. D. W. McKee, *J. Catal.* **14**, 355 (1969).
589. L. H. Slaugh and J. A. Leonard, *J. Catal.* **13**, 385 (1969).
590. Asahi Kasei K.K.K., Brit. Pat., 1,110,687 (1968); *Platinum Metals Rev.* **12**, 119 (1968).
591. L. H. Little, "Infrared Spectra of Adsorbed Species." Academic Pres, New York, 1966.
592. G. C. Bond, T. J. Philipson, P. B. Wells, and J. M. Winterbottom, *Trans. Faraday Soc.* **62**, 443 (1966).
593. S. Siegel, *Advan. Catal.* **16**, 123 (1966).
594. T. Krück, K. Baur, and W. Lang, *Chem. Ber.* **101**, 138 (1968).
595. J. Chatt, F. A. Hart, and D. T. Rosevear, *J. Chem. Soc., London* p. 5504 (1961).
596. E. O. Fischer and G. Burger, *Z. Naturforsch. B* **16**, 702 (1961).

Appendix

A number of papers pertinent to the topics discussed here appeared during the preparation of this volume, but too late for inclusion into the main text; they are briefly summarized here. The page numbers quoted are to the relevant discussions in the text.

R. P. Hughes and J. Powell [*Chem. Commun.* p. 275 (1971)] have shown that a dynamic σ-allyl is formed in the reaction of dienes with π-allylic complexes (pp. 41, 42) and that the rate-determining step of the overall reaction is insertion into the Pd-σ-allyl bond, not complexation of the diene, nor formation of the σ-allyl complex. D. Medema and R. van Helden [*Rec. Trav. Chim. Pays-Bas* **90**, 304, 324 (1971)] have given some further details of their work on this reaction (p. 45). They also gave evidence for the structures of intermediate complexes and for a reaction mechanism involving binuclear intermediates in the di-, tri-, and tetramerization of butadiene by allylic palladium acetate catalysts.

K. E. Atkins, W. E. Walker, and R. M. Manyik [*Chem. Commun.* p. 330 (1971)] and T. Mitsuyasu, M. Hara, and J. Tsuji [*Chem. Commun.* p. 345 (1971)] have reported the formation of hydroxyoctadienes and octadienylamines from butadiene and water or ammonia (p. 43). The catalysts used were palladium acetylacetonate and palladium acetate, respectively, both in the presence of triphenylphosphine. K. Suga, S. Watanabe, and K. Hijikata, [*Aust. J. Chem.* **24**, 197 (1971)] have reported the acetoxylation and dimerization of isoprene (pp. 43, 105).

The cyclodimerization of butadiene with phenyl isocyanate to give substituted *N*-phenylpiperidones has been reported by K. Ohno and J. Tsuji [*Chem. Commun.* p. 247 (1971)] (p. 46).

Arylthallium(III) compounds react with $PdCl_2$ and NaOAc in acetic acid to give biaryls (p. 65) according to S. Uemura, Y. Ikeda, and K. Ichikawa [*Chem. Commun.* p. 390 (1971)].

Moritani and his coworkers have proposed that the arylation of olefins proceeds via a σ-vinyl σ-aryl complex, rather than as suggested on pp. 69–70 [I. Moritani, Y. Fujiwara, and S. Danno, *J. Organometal. Chem.* **27**, 279 (1971)]. Evidence in support of this proposal comes from the observation that σ-vinyl(chloro)bis(triphenylphosphine)palladium(II) complexes react with benzene in the presence of silver acetate to give styrenes, for example,

$$trans\text{-}(Ph_3P)_2Pd(Cl)CH{=}CHCl + C_6H_6 \xrightarrow{AgOAc} trans\text{-}PhCH{=}CHCl \quad (82\%)$$

Both the *cis*- and *trans*-β-chlorovinyl complexes gave *trans*-β-chlorostyrene. While the mechanism which is implied by this result is certainly feasible, it is by no means clear that the reaction as normally carried out (using the free olefin and in the absence of triphenylphosphine) actually proceeds by this path. There is some evidence to support the possibility of an electrophilic palladation of an olefin (p. 63) but this would appear to be an uncommon process by comparison to the more normal insertion reactions. For this mechanism mono-alkyl olefins and vinyl ethers should be more reactive than other olefins but the evidence available[235,237,238] suggests this not to be so.

T. Mizoroki, K. Mori, and A. Ozaki [*Bull. Chem. Soc. Jap.* **44**, 581 (1971)] have reported a variant on the above reaction in which an aryl iodide is used,

$$PHI + CH_2{=}CHR + KOAc \xrightarrow{PdCl_2} PhCH{=}CHR + HOAc + KI$$

Catalytic amounts of $PdCl_2$ are used but metallic palladium, which is also an active catalyst, is precipitated during the reaction.

Some details of the conversion of ethylene to 2-chloroethanol (pp. 80, 155 and reference 279) have been reported by H. Stangl and R. Jira [*Tetrahedron Lett.* p. 3589 (1971)], who obtained good yields at high ethylene pressures and high chloride ion concentrations. Copper chloride was necessary as a cocatalyst and the reaction could be run catalytically in the presence of oxygen.

P. M. Henry and G. A. Ward [*J. Amer. Chem. Soc.* **93**, 1494 (1971)] have analyzed the products obtained from the reaction of cyclohexene-3,3,6,6-d_4 with Pd(II)/$CuCl_2$/LiOAc in acetic acid. The first step proposed is an acetoxypalladation, which can then be followed by an isomerization. The products are generated either by elimination of H–Pd (to give the unsaturated esters) or by reaction with $CuCl_2$ to give the saturated chloroesters. From the distribution of deuterium in the products the authors concluded that acetoxypalladation of the double bond occurs trans while elimination of H–Pd is cis.

The same reaction, but under rather different conditions (nitrogen oxides as cocatalyst but no copper or halide present), was studied by S. G. Wolfe and P. G. C. Campbell [*J. Amer. Chem. Soc.* **93**, 1497 and 1499 (1971)]. In this case only unsaturated acetates were obtained; the allylic acetates were suggested to arise from a π-allylic intermediate, and two mechanisms, one similar to that of Henry and Ward, were offered to account for the homoallylic acetates. These authors also examined the behavior of some π-cyclohexene and π-cyclohexenyl complexes as models for this reaction.

P. M. Henry [*Chem. Commun.* p. 328 (1971)] has reported that two paths coexist for the catalyzed isomerization of allylic esters (see also pp. 110, 126). One, which occurs with exchange, involves an oxypalladation step, while the other, which does not involve exchange of the ester, occurs via a σ-bonded cyclic acetoxonium ion.

The acetoxylation of substituted benzenes by palladium acetate in the presence of oxygen has been shown by L. Eberson and L. Gomez-Gonzales [*Chem. Commun.* p. 263 (1971)] to give largely the meta-substituted products. Under these conditions little side-chain acetoxylation occurred (see p. 118).

M. Graziani, P. Uguagliati, and G. Carturan [*J. Organometal. Chem.* **27**, 275 (1971)] found that ethanol was carbonylated in the presence of $PdCl_2$ and LiCl to give ethyl chlorocarbonate and ethyl acetate. In the presence of base only diethyl carbonate was formed (p. 126).

The Pd(0) catalyzed hydrosilation of terminal olefins by Cl_3SiH (p. 162) has been reported by M. Hara, K. Ohno, and J. Tsuji [*Chem. Commun.* 247 (1971)].

F. Goodridge and C. J. H. King [*Trans. Faraday Soc.* **66**, 2889 (1970)] have shown that ethylene was efficiently electrochemically oxidized to acetaldehyde at a Pd electrode in dilute sulfuric acid (see also p. 167).

Author Index

Numbers in parentheses are reference numbers and indicate that an author's work is referred to, although his name is not cited in the text. Numbers in italics show the page on which the complete reference is listed.

C

D

I

J

K

L

M

N

T

U

V

W

Y

Z

Index of Reactions

The major palladium-catalyzed reactions are listed here; for further details see the Subject Index.

R

S

T

V

Subject Index

C

D

N

O

P

R

S

T